The Matrix

The Matrix

Charting an Ethics of Inheritable Genetic Modification

Marilyn E. Coors

ROWMAN & LITTLEFIELD PUBLISHERS, INC.
Lanham • Boulder • New York • Oxford

ROWMAN & LITTLEFIELD PUBLISHERS, INC.

Published in the United States of America
by Rowman & Littlefield Publishers, Inc.
A Member of the Rowman & Littlefield Publishing Group
4720 Boston Way, Lanham, Maryland 20706
www.rowmanlittlefield.com

12 Hid's Copse Road
Cumnor Hill, Oxford OX2 9JJ, England

British Library Cataloguing in Publication Information Available

Library of Congress Cataloging-in-Publication Data

Coors, Marilyn E., 1947–
The matrix : charting an ethics of inheritable genetic modification / Marilyn E. Coors.
p. cm.
Includes bibliographical references and index.
ISBN 0-7425-1400-5 (cloth : alk. paper) — ISBN 0-7425-1401-3 (pbk. : alk. paper)
1. Genetic engineering—Moral and ethical aspects. 2. Biotechnology—Moral and ethical aspects. I. Title.
QH438.7 .C665 2002
174'.966—dc21

2002005245

∞™ The paper used in this publication meets the minimum requirements of American National Standard for Information Sciences—Permanence of Paper for Printed Library Materials, ANSI/NISO Z39.48-1992.

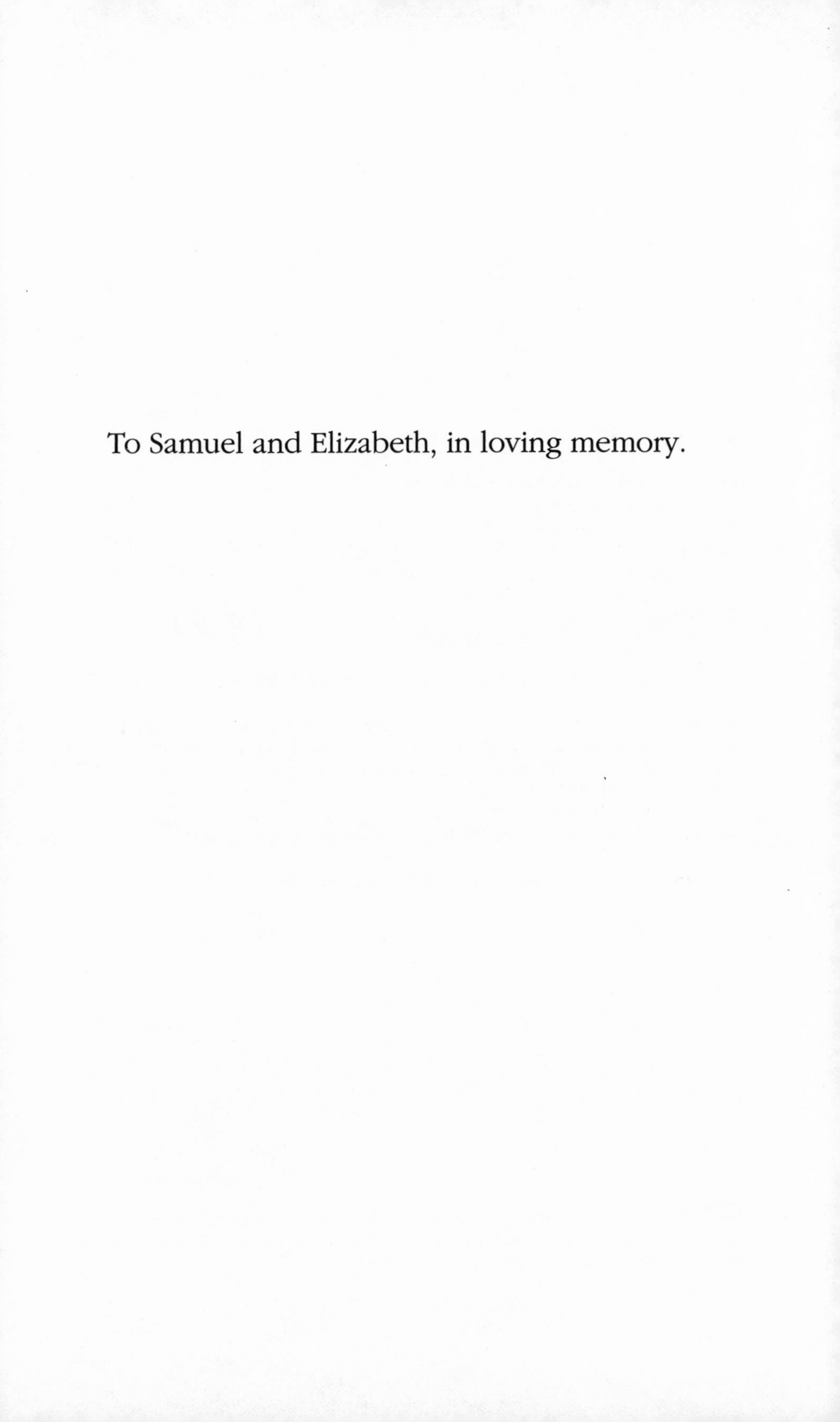

To Samuel and Elizabeth, in loving memory.

Contents

Acknowledgments

Many individuals shared in the journey that led to the publication of this book. I would like to thank them all, but that is impossible.

I owe a great debt of gratitude to faculty in the departments of philosophy and religion in the joint doctoral program at the University of Denver: William Gravely, Dana Wilbanks, Naomi Reshotko, Jere Surber, and Edward Everding. I would also like to acknowledge Mary Ann Cutter from the University of Colorado, Colorado Springs, for contributing her expertise in bioethics. I thank all of these colleagues for their scholarly input and encouragement throughout the writing and publishing process.

Special thanks go also to my colleagues, Mark Yarborough and Therese Jones, in the Center for Bioethics and Humanities at the University of Colorado Health Sciences Center for their day-to-day interest and feedback on this project.

I am especially grateful to Jeannie Zuk, R.N., for her assistance with the initial and most difficult proofreading of this book.

My heartfelt appreciation goes to editor James Langford who wanted to publish this book from the first time he read the proposal. He has never wavered in his enthusiasm for this publication.

Geneticists, Carol Green, M.D., and Peter Lane, M.D., deserve accolades for returning my phone calls and providing much-needed information on the science and ethics of sickle cell disease.

And, above all, I would like to thank my husband, Peter, and children, Melissa, Christien, Peter, Carrie, Ashley, and David, for their understanding and support during periods when I was preoccupied and unavailable. It is because of them, their children, and future generations that we must chart an ethical approach to inheritable genetic modification.

> If we do not think about it now, the possibility of our having a free choice will, one day, suddenly be gone.
>
> —James D. Watson, 1971, before Congress

Introduction

"Alexander the Great once remarked that the peoples of Asia were slaves because they had not learned to pronounce the word 'No!'"[1] Sir Winston Churchill used this quotation from antiquity as a cautionary message for a world in crisis, to challenge the corrosion of international will that was evident in the mollification of Adolf Hitler. As the title of this book implies, a "no" is appropriate for certain uses of inheritable genetic modification, which is intentional changes to germline cells (egg or sperm) or early embryos, thereby altering both the immediate person and his or her descendants. Extraordinary power such as this has not only beneficent features, but maleficent ones as well. Nations or peoples are great when they know when to say "no" or "yes" to powerful leaders or technologies. The ideals that empower a society or culture are operational when they can both sanction and prohibit endeavors.

We are experiencing a profound crisis of authority in contemporary America, which is reflected in the blurring of beliefs and boundaries at the highest echelons. This is the consequence of dissimilar worldviews and divergent conceptions of moral authority, resulting in ongoing disagreement over standards and norms. This crisis is manifest in indistinct convictions, disintegrating boundaries, and culture wars that are witnessed at all levels of society, and it culminates in the erosion of the significance of values and the implausibility of determining shared beliefs upon which to establish a moral mooring. The plethora of alternatives produces clashing ideas concerning the meaning of human existence, both individually and collectively.

The diminution of moral authority occurred not because of, but concurrent with, the development of human genetic technology. The capacity to alter the human genome, however, initiated changes in perceptions, actions, and consequences on a monumental scale. Developments in molecular genetics prompted philosophers and ethicists to attest to the inadequacy of former ethical systems. Thus, just when the need for moral direction is possibly at an all-time high, the basis of moral authority faces a large question mark. Professionals and laypeople alike are forced to navigate a moral domain fraught with a plurality of virtues, values, and principles and a dearth of moral resources to use as a rudder.

We are experiencing the genesis of inheritable genetic modification in humans. The challenges posed by the potential of this technology are of a magnitude that many believe is unprecedented in human history. The initial course of inheritable genetic modification could forebode or decree what is to come. Will this new genetic technique deliver on its promise to cure intractable genetic disorders for many, or will it bestow advantageous characteristics upon only a few? Will it improve the human condition or merely further commodify life? The values that society chooses to guide the launch of inheritable genetic modification will determine whether it ultimately benefits or harms humankind.

In response, the aim of this work is to engage ethics, religion, and science in an interdisciplinary dialogue to identify pertinent virtues, duties, and principles that differentiate the beneficial uses of inheritable genetic modification from those that threaten the dignity of human life. It culminates in an ethical perspective that is summarized in the Inheritable Modifications Matrix (IMM), which is intended to guide moral deliberations concerning the benefits and harms involved.

Three main points underlie the thrust of this effort: First, biotechnology is a force so powerful that it calls for an expanded ethical perspective to guide its use. Biotechnology has the potential to alter the primordial material of human life in a way that has not been possible until now. It enables humans to function as cocreators rather than as stewards of creation, straining the role of human predictive wisdom in an unprecedented fashion. Second, the challenges of inheritable genetic modification warrant an alliance of disciplines and ethical theories to respond comprehensively to the possibilities of this new technique. Few moral agents exclusively utilize one ethical theory in responding to complex dilemmas; therefore, given the intricacy of the issue, an inclusive approach to ethics is necessary. Third, I propose the IMM as an ethical perspective with the breadth to address the questions that this technology poses. The matrix juxtaposes pertinent virtues, duties, and ethical principles with the scientific aspects of inheritable genetic modification. The intent of the matrix is to facilitate ethical deliberation and assist moral decision-making processes. The goal is to avert disputes stemming from divergent worldviews and to engage persons of differing beliefs by bypassing peripheral issues that are likely to divide them.

The framework of the moral event provides the infrastructure for the matrix. The moral event is comprised of the four components: "the agent, the act, the circumstance, and the consequences—in relation to each other."[2] The advantage of this structure is that it permits the interrelationship of a variety of disciplines, each clarifying important aspects of moral judgment. The interplay of philosophy, religion, and science differs, depending upon the facet of the moral event considered. For example, philosophy and religion contribute more heavily to the consideration of the agent, the action, and the circumstances, with science playing a more substantive role in analyzing the consequences. The underlying premise of this book is that any one ethical theory, alone, is inadequate for dealing with the challenges of this new technology. Inheritable genetic modification raises distinct ethical issues, for it brings about purposeful and specific changes in individual genes that are later introduced into the gene pool. This is a departure from any former medical procedure, since it entails intergenerational consequences that were heretofore unfathomable.

This premise will not elicit an in-depth analysis of the respective ethical theories that are considered. I am fully aware of the limitations of treating comprehensive theories with a broad brush. Nonetheless, I maintain that even a more nuanced consideration of each reveals insufficiencies when called upon to address the challenges of inheritable genetic modification. For that reason, I will align pertinent contributions from four comprehensive theories to create an ethical framework that can assess inheritable genetic modification. The virtue theory of contemporary physician and philosopher Edmund Pellegrino focuses on the agent and his or her character. The Imperative of Responsibility for the technological age, proposed by twentieth-century philosopher Hans Jonas, informs the discussion of the act, and whether it is right or good, in keeping with moral duty. The particularistic ethic of Emmanuel Levinas represents the philosophical contribution of postmodernism to the analysis of the moral event. Levinas's work contributes to the definition of ethics and to the discussion of the individual, communal, and historical circumstances that define each unique ethical agent. Finally, the philosophy of utilitarian ethicist John Stuart Mill informs the consideration of the consequences of the act as they enhance or undermine human well-being.

This book is divided into three parts. Part I reviews the scientific and ethical issues involved with this technology to provide context for the matrix. Part II addresses the four components of the moral event to identify the operative virtues, duties, and principles that form the matrix's ethical axis. Part III presents the Inheritable Modifications Matrix, which is intended to stimulate and guide ethical reflection concerning inherited changes to the human genome, and proposes an ethical boundary for the technology.

Throughout history, people have displayed an enthusiasm for unlimited creativity in human endeavor. Their common motive is to gain the liberty to seek one's own truth free from restraints. Inheritable genetic modification offers us the opportunity to test the limits of human knowledge and predictive wisdom. The following investigation demonstrates that it is necessary to restrain this almost irresistible enticement and pause a moment for moral deliberation in the process.

NOTES

1. Winston Churchill, in Os Guinness, "The Wise Art of Giving" (Burke, Va.: Trinity Forum Seminar Curricula, The Trinity Forum, 1996).

2. Edmund Pellegrino, "Toward a Virtue-Based Normative Ethics for the Health Professions," *Kennedy Institute of Ethics Journal* 5, no. 3 (September 1995), 253–277.

1

THE ETHICS OF HUMAN GENETIC MODIFICATION

Nineteenth-century author Nathaniel Hawthorne used fiction to illustrate humankind's quest for great power over nature. He frequently depicted the physician-scientist working in isolation to epitomize unethical experimentation with human subjects. In his short story, *The Birthmark,* Hawthorne portrays a fanatical scientist, Aylmer, who is obsessed with his beautiful wife's single physical flaw, a crimson birthmark on her otherwise flawless cheek. "Georgiana, you came so nearly perfect from the hand of nature that this slightest possible defect, which we hesitate whether to term a defect or a beauty, shocks me, as being the visible mark of earthly imperfection."[1] In his fervor to erase the despised mark and render his wife perfect, Aylmer creates a powerful potion in his laboratory. His devoted spouse, presumably worn down by his disdainful gaze, consents to drink the potion: "Danger is nothing to me; for life, with this hateful mark makes me the object of your horror and disgust—life is a burden which I would fling down with joy."[2] Aylmer begins the experiment completely confident of its success. It results in the simultaneous disappearance of the despised birthmark and the death of Georgiana.

Although the experiment is dated, *The Birthmark* raises many timely issues. The ephemeral ideal of human perfection is still paramount, and defects are often considered insufferable. Flawlessness is necessary in order to qualify for full acceptance in a relationship or in society. Also, the dangers of unfettered research are a recurring theme. The reader condemns Aylmer and his secret potion by virtue of its consequences, but more importantly, because his desire for power and perfection was undaunted by ethical mores and devoid of a sense of responsibility toward his research subject. Moreover, the moral of the story convicts Aylmer. Obsessed by aspirations of scientific prowess, he fails to recognize the true beauty of human life.

NOTES

1. Nathaniel Hawthorne, "The Birthmark," in *The Portable Hawthorne,* ed. Malcolm Cowley (New York: Viking Press, 1948), 164–185.
2. Hawthorne, "The Birthmark," 169.

1

The Science

Any science is of contraries—of the object and its opposites: if of good, then also of evil, if of building, then also of destroying, if of health, then also of disease, if of life, then also of death: from which alone it would seem to follow that only in the hands of angels would the power of science be sure to be for the good only.

—Hans Jonas, *The Imperative of Responsibility*

The history of the Human Genome Project (HGP), a $3 billion international effort to locate, map, and sequence the DNA of approximately 35,000 to 50,000 human genes, contrasts Aylmer's approach in many ways. From its inception, the HGP was possibly the most scrutinized scientific undertaking in history. It began with a self-imposed yearlong moratorium to study the ramifications of DNA research in order to ensure that the field would develop safely; only after careful scrutiny, an interdisciplinary team decided that it was prudent to resume the research. In addition, most of the participating nations allocated a part of their budget for the project to tackle the ethical, legal, and social issues involved with human genetics. Yet, the story of DNA research is incomplete if its similarities to Aylmer's endeavors go unmentioned. Its allure is seductive, powerful, and not without safety risks or moral concerns. Georgiana expressed it marvelously: "I know not what may be the cost to both of us to rid me of this fatal birthmark. Perhaps its removal may cause cureless deformity; or it may be the stain goes as deep as life itself."[1] Each of us must ask the same questions of the new genetics: What will it cost? Will it go as deep as life itself?

In order for nonscientists to consider these questions and participate in the public debate, it is first necessary for them to understand the basic concepts of human genetics. Science tells us a great deal about the molecular processes that control the development and maintenance of life, but cannot tell us what is morally right or wrong for humans. In other words, it tells us what we *can* do, but not what we *should* do. Conversely, a bioethical analysis that ignores or distorts pertinent science is either uninformed or disingenuous. Let us, therefore, have a brief overview of the science before delving into the ethical concerns.

THE SCIENCE

The human genome is the aggregate of genetic material contained in the twenty-three pairs of chromosomes in each human cell; some refer to the genome as the "book of life." A gene is the basic functional unit of inheritance, and is to heredity what a paragraph is to a book. Genes consist of the macromolecule deoxyribonucleic acid (DNA), which is comprised of four nucleotide bases, a sugar (deoxyribose), and a phosphate. The nucleotide bases are analogous to letters in the alphabet, for they are to the genetic code what letters are to written language: a linear sequence of nucleotide bases comprises a gene, just as a sequence of letters denotes a word. The many combinations made possible by these four elemental letters of the genetic alphabet (A, G, C, and T) are the source of the marvelous complexity that resides in DNA. To continue the analogy, a different arrangement of the same letters can produce a different gene or word. For example, transposition of the letters in the verb "pare" produces the noun "pear." These two words have different grammatical functions and unrelated meanings, and interchanging them will disrupt the meaning of a sentence.

Every gene provides information that instructs cells to produce the proteins necessary for life. The normal functioning of all genes in concert, each producing a precise amount of protein at specific times and in exact locations, is necessary in order to develop and maintain life. When a deletion, insertion, repetition, or transposition of bases occurs in a segment of DNA that manufactures protein, the cell produces a nonfunctional protein, no protein at all, or an incorrect amount. A DNA miscoding sequence resembles "typing mistakes that would be expected of a secretary transcribing a biochemical manuscript some 3 billion characters long."[2] The outcome of a change in genes, which are the coding regions of DNA, may be devastating disease, while a change in the noncoding regions of DNA may have no effect at all.

Some examples of disorders caused by a DNA mutation in a coding region are sickle cell anemia, cystic fibrosis, Huntington's disease, Type 2 diabetes, and many cancers. However, isolating a dysfunctional gene to correct these diseases compares to determining the location of a burned-out lightbulb in a home located somewhere in the United States. It is necessary to have road maps with markers before attempting such a task. The HGP completed a genetic map that includes more than 2,000 molecular markers and variable DNA sequences that serve as landmarks for locating genes.[3] The challenge of the postgenome era is to determine the function of genes, particularly how they work in concert with the environment to promote health or disease.

One of the many endeavors of the postgenome era is the attempt to repair or replace dysfunctional genes linked to disorders. The technique used to correct genetic errors or to enhance normally functioning genes is genetic modification, which is possible with two types of cells: 1) somatic (body) cells, or all cells lacking the potential to become reproductive cells; and 2) germ line cells, which are reproductive cells (eggs or sperm) or early embryos. Intentional changes to germ line cells are called inheritable genetic modification. The technical obstacles involved in changing genes safely and effectively are daunting, as will be discussed below.

SOMATIC CELL GENE MODIFICATION

Somatic cell gene modification is the transfer of a gene encoding for a specific substance into a body cell to restore health or enhance function. Genetically altered somatic cells convey the transferred gene to daughter cells upon cell division, but do not pass on changes to future generations. W. French Anderson, a human geneticist and pioneer of gene therapy, performed the first successful somatic cell gene therapy in 1990. Two sisters with a rare genetic disease, adenosine deaminase (ADA) deficiency, received white blood cells carrying a functional ADA gene. Anderson hoped that the transferred gene would produce the amount of ADA necessary to enable their immune systems to operate properly. Despite the fact that the girls did well clinically, there is no unambiguous evidence that gene therapy deserves the credit. The children were also given regular injections of synthetic ADA, a conventional treatment that may instead be responsible for their good health. Anderson used the results of this trial to document that "all the right things are happening to insure gene therapy's ultimate success," yet admitted that "gene therapy is not yet curing patients."[4]

In September 1999, Jesse Gelsinger, an eighteen-year-old patient with ornithine transcarbamylase (OTC) deficiency, died as a direct result of experimental gene therapy at the University of Pennsylvania. OTC is an inborn error of metabolism that compromises liver function. When Gelsinger entered the trial, his liver function met the criteria for acceptance into the research study. However, laboratory tests revealed that his liver function was below the required criteria when the genes were actually infused. Within hours of receiving the gene infusion, which was encased in a viral transport mechanism (gene therapy often uses modified viruses to transport therapeutic genes into cells), Gelsinger's immune system went out of control. Four days later, he died of multiple organ failure. In this case, researchers suspect that his immune system triggered an acute inflammatory response to the virus.[5] Toxicity of the viral transport mechanism used in the Gelsinger case has also been documented in preclinical testing of gene therapy for hemophilia B in monkeys.[6]

The charges of inappropriate conduct by the scientists responsible for the Gelsinger case included neglect of safety checks, disregard for established practices, and unreported serious adverse events in animal and human trials using viral vectors. The investigation also concluded that the informed consent process failed to inform the patient and his parents of the risks associated with the procedure. In addition, this tragic occurrence precipitated an intense review of technical and compliance problems involved with somatic cell gene therapy and caused significant damage to public trust in biomedical research.[7] Serious adverse events in early phases of a study do not necessarily doom a procedure to failure, since it is impossible to conduct research in a zero-risk environment. Unfortunately, patients will suffer and die even when all criteria meant to ferret out danger are strictly enforced. This is the price that society generally accepts in order to realize the tremendous benefits of new drugs and therapies. That said,

Gelsinger's untimely death underscores the obligation to understand why early attempts at gene therapy failed to produce anticipated results and to protect future research subjects against these and other harms.

Recently, investigators announced exciting new advances in somatic cell gene therapy that improved the health of patients suffering from severe combined immunodeficiency (SCID), a rare and lethal immune disease sometimes called the "bubble boy disease." A normal version of the immune system gene was added to a "harmless" virus and introduced into the patients' cells.[8] Early results confirm that their immune systems are functioning well, without any signs of harmful side effects. Similar results show promise in trials with hemophilia B, a genetically inherited bleeding disorder,[9] and cystic fibrosis, a fatal pulmonary disorder.[10] These research trials also continue to work on a transvection technology (gene transport device) to improve efficiency and avoid the toxic effect witnessed in the Gelsinger case. Prospects for the success of somatic cell gene therapy continue to improve, and some feel that the success of new therapies has now moved from the realm of the possible to the probable.

Assuming that scientists overcome the tremendous challenges of safety, expression, and stability in gene therapy, somatic cell gene alteration is an extension of standard medical practice and is not significantly different from organ donation, blood transfusion, or many other medical procedures. For example, in organ transplantation, the patient receives a functional organ from an outside source to replace his or her dysfunctional organ. The procedure involves the informed consent of the party involved, and the effect of the intervention ceases to exist when the patient dies. Organ transplants result in significant biological differences for the post-treatment patient, just as somatic cell gene alteration does. Also, in both procedures, the future existence of the patient is contingent upon technological intervention. Both organ transplantation and somatic cell genetic alterations are ethically justifiable when they comply with the principles of autonomy (informed consent), beneficence (do good and avoid harm), and justice. The present ethical domain encompasses these medical interventions because traditional bioethical theories focus on caring for and treating disease among existing persons. If the scope of genetic alteration were confined to their somatic cell changes and did not affect those of future generations, it would present no unique ethical challenges.

INHERITABLE GENETIC MODIFICATION

Inheritable genetic modification, as discussed earlier, is the transfer of a functional gene or genes into germline cells or early embryos to purposefully modify every cell of the future person. It is done via a viral transport mechanism that injects a gene at a random location in the host genome in much the same way that a cold virus infects cells. Geneticists also manipulate the germline by introducing a new additional chromosome carrying a functional gene (or genes) into cells. This approach is just becoming feasible. It is an important advance since it leaves the original genome untouched and since there is evidence that artificial chromosomes will persist for repeated divisions when introduced into human cells.[11] The technology for safe and effective inheritable genetic modification is not yet available.

Scientific advances in animal models foretell ongoing progress in this arena. Researchers developed a method in sheep that can deliver a gene into, or remove it from, a particular location.[12] This improvement, known as gene surgery, has the ability to pinpoint precise genome sites for alteration. This technique is a significant achievement in that genes inserted at a specific site (rather than at a random location) in the genome avoid the potential to disrupt other correctly functioning genes. This procedure is not yet perfected, but preliminary results indicate dramatic advances beyond what scientists are able to do presently with other types of gene therapy. Despite differences in complexity between animals and humans, developmental similarities indicate that comparable results could occur in humans.

The first evidence of inheritable genetic modification that resulted in normal healthy children was reported in 2001.[13] A procedure, which to date is responsible for the birth of thirty babies worldwide, adds the inner liquid of a donor egg to the egg of an infertile woman to reactivate it. This process results in a child with mitochondrial DNA from two different women. Mitochondria are organelles in the inner liquid of the egg that contain a few genes essential for life. Mitochondrial DNA is the only DNA located outside of the cell's nucleus, and it is inherited only from the mother. Mitochondrial DNA affects the expression of a cell's nuclear DNA and can have a range of physiological effects on the resulting child. Both male and female children that are produced by this method inherit the new mitochondria. However, in a female child the new DNA is later passed on to her children via her eggs, constituting an inheritable genetic modification. The success of this procedure may bring inheritable changes of nuclear DNA a step closer.

A second instance of inadvertent inheritable genetic modification was reported in a gene therapy trial that used a virus to transfer a gene, which later appeared in the semen of a sixty-three-year-old patient with severe hemophilia.[14] The man received a low dose of a virus carrying a therapeutic gene for blood clotting factor IX into his liver, and subsequent tests on his semen detected gene sequences from the virus, and his semen tested positive for the virus for seven weeks. The detection of the viral gene fragments in the semen does not verify that the virus actually entered the sperm, since semen also contains white blood cells and other types of cells. However, the Food and Drug Administration (FDA) halted the trial even though the patient's sperm tested negative for the gene fragments. Additional patients were allowed to enroll in the study only after their semen tested negative for three months, since that is the time it takes for sperm to mature. This study highlights the possibility that gene therapy may unintentionally result in inheritable genetic modifications that affect both the patient and his children.[15] This technique is ethically different from standard medical procedures in that precise changes are permanently encoded in the genome of future persons and potentially become part of the human gene pool. The Concerns and Justifications section (below) introduces the ethical reactions to inheritable genetic modification.

Therapy and Enhancement

The possible applications of genetic technology are sometimes distinguished according to purpose: therapy or enhancement. This distinction attempts to differentiate between treatments intended to prevent or cure disease (therapy) and those meant to improve function or appearance beyond what is considered "normal" for members of our species (enhancement). The latter targets characteristics like physical function, intelligence, appearance, strength, size, and stature, with the intent of creating more desirable qualities than would be the result of natural reproduction.

The dividing line between therapy and enhancement is not always clear-cut. For example, Huntington's disease is a late onset fatal neurodegenerative disorder, so a change to correct the Huntington's gene is clearly therapeutic. However, alterations to genes that predispose individuals to common complex disorders, such as heart disease or autoimmune disorders, are less clear. It is possible that genetic alteration techniques could someday eliminate a predisposition to heart disease, thereby increasing life expectancy by several decades. Is this therapy or enhancement? Moreover, does an alteration to the autoimmune gene(s) that enables an individual to function under conditions that are dangerous for others qualify as therapy or enhancement? Presently, heath insurance companies attempt to distinguish between therapy and enhancement based on the medical benefit of the treatment and the presence of an "inherited condition" versus an "underlying condition." While there are no specific definitions for these terms, some examples might help to clarify the distinction. A typical company will cover surgical reconstruction for a child born with a cleft lip based on the designation of "inherited condition" and pay for speech therapy following surgery for up to twenty sessions per year. Once the child's speech ceases to improve, the company discontinues coverage of further therapy because the "underlying condition" precludes medical benefit. Similarly, in the case of a short-statured child, the company will pay for treatment based on a diagnosed growth hormone deficiency and the predicted success rate of the therapy. Once marked improvement in height comes to an end, the company typically ceases coverage of the treatment. The questions surrounding hormone therapy that enhances height above "normal" remain debatable. What is considered "normal"? Where is the line in the alteration of height—from 4'4" to 5'4" or even 6'4" (highly unlikely)? At what point does therapy become enhancement? The difficulties with this distinction are obvious. There is a natural progression from obviously therapeutic to clearly enhancement purposes, leaving a gray area that is difficult to classify. As science links more and more genes to disorders, the cases that blur this distinction will only increase.

Concerns and Justifications

The scientific and ethical complexities inherent in the evaluation of inheritable genetic modification are numerous. This technique promises tremendous benefit for those who suffer from intractable genetic disorders, but it also presents the potential for extraordinary control over human biology and personality. It has the potential to modify traits that are essential to our humanity such as reason, moral choice, or emotion. The mandate for public deliberation concerning the benefits and risks of this new technology appears in professional journals and the media regularly. The American Association for the Advancement of Science (AAAS) put together an interdisciplinary task force composed of lawyers, theologians, ethicists, and scientists to discuss inheritable genetic modification and offer recommendations.[16] The group called for a moratorium on inheritable genetic modification and on current therapies that cause inheritable modification inadvertently, including (as mentioned earlier) injecting infertile women's eggs with the interior liquid of donor eggs. This moratorium is based upon existing technical obstacles to safe and effective gene correction or replacement. Because these changes are transmitted to future generations, compelling evidence of their safety and efficacy must be demonstrated before they are considered appropriate for human use. Also, inheritable genetic modification entails pressing moral concerns: its full impact on our attitudes about humans, social justice, the nature of human reproduction, and parent-child relations needs further consideration before embarking on this course. The AAAS task force recommended that human trials of inheritable genetic modification should be curtailed until safe and reliable techniques for genetic change or replacement are developed. The report recommends that if society decides to proceed with inheritable genetic modification, extensive oversight mechanisms should be created to review and approve all trials.

The AAAS also cites justifiable medical and ethical reasons to consider the utilization of inheritable genetic modification once it is perfected. The first is medical necessity, which includes the obligation to rectify developmental disorders and those disorders that affect multiple systems but are not amenable to somatic cell treatment. Inheritable genetic modification could remedy these otherwise irreversible afflictions permanently. The second validation is medical efficiency. Inheritable genetic modification is more efficient than the repetitive somatic cell gene alteration over successive generations because somatic cell modification involves its own risks, as discussed above. Thus, it would be safer and more cost-effective to eliminate a disorder in both the present and future generations in one procedure. Third, inheritable genetic modification eliminates the need for repeated prenatal diagnostic testing and selective abortion in families afflicted with genetic disease. With advances in safety and efficacy, inheritable genetic modification may provide a justifiable option, technically as well as morally, for those who oppose abortion. Fourth, when both parents are homozygous for the same genetic disorder, screening and selection procedures are not an option. Inheritable genetic modification offers the potential to prevent genetic disease in their offspring. Finally, the medical profession should pursue and implement the best available techniques to prevent or treat genetic disease whenever possible.

The ethical issues generated by inheritable genetic modification in humans are numerous and profound. They form the subject matter of the following chapters. For the purpose of our discussion, these issues include bioethical, religious, and philosophical concerns. This is not to indicate that the line between the groupings is impenetrable or categorical. What I want to pursue in this book is an interdisciplinary dialogue that juxtaposes science, philosophy, and religion to address the challenges posed by inheritable genetic modification. The following three chapters examine diverse perspectives in an effort to establish the groundwork for the discussion of its ethical boundary.

NOTES

1. Nathaniel Hawthorne, "The Birthmark," in *The Portable Hawthorne,* ed. Malcolm Cowley (New York: Viking Press, 1948), 169.

2. Constance Engelking, "The Human Genome Project Exposed: A Glimpse of Promise, Predicament, and Impact on Practice," *Oncology Nursing Forum (Supplement)* 22 (1995): 27–34.

3. See "The Human Genome," *Nature* 409 (15 January 2001): 745–964; and "The Human Genome," *Science* 291 (17 February 2001): 1145–1434.

4. W. French Anderson, "End-of-the-Year Potpourri—1996," *Human Gene Therapy* 7 (1996): 2201–2202.

5. Adam Bostanci, "Blood Test Flags Agent in Death of Penn Subject," *Science* 295 (25 January 2002): 604–605.

6. Jay Lozier, G. Csako, T. Mondaro, D. Krizek, M. Metzger, R. Costello, J. Vostal, M. Rich, R. Donahue, and R. Morgan, "Toxicity of a First Generation Adenoviral Vector in Rhesus Macaques," *Human Gene Therapy* 13 (1 January 2002): 113–124.

7. Theodore Friedman, "Principles for Human Gene Therapy Studies," *Science* 287 (24 March 2000): 2163–2168.

8. Marina Cavazzana-Calvo, Salima Hacein-Bey, Genevieve de Saint Basile, Fabian Gross, Eric Yvon, Patrick Nusbaum, Francoise Selz, Christophe Hue, Stephanie Certain, Jean-Laurent Casanova, Philippe Bousso, Francoise Le Deist, and Alain Fischer, "Gene Therapy of Human Severe Combined Immunodeficiency (SCID)-X1 Disease," *Science* 288 (2000): 669–672.

9. K. A. High, "AAV-Mediated Gene Transfer for Hemophilia," *Annals of the New York Academy of Science* 953 (December 2001): 64–74.

10. T. Float and B. Laube, "Gene Therapy in Cystic Fibrosis," *Chest* 120 (September 2001): 124S–142S.

11. Salima Hacein-Bey-Abina, Françoise Le Deist, Frederique Carlier, Cecile Bouneadu, Christophe Hue, Jean-Pierre De Villartay, Adrian J. Thrasher, Nicolas Wulffraat, Ricardo Sorensen, Sophie Dupuis-Girod, Alain Fischer, Marina Cavazzana-Calvo, "Sustained Correction of X-Linked Severe Combined Immunodeficiency by Ex-vivo Gene Therapy," *The New England Journal of Medicine* 346 (18 April 2002): 1185.

12. J. McCreath, K. H. S. Howcroft, A. Campbell, A. Coleman, A. E. Schnieke, and A. J. Kind, "Production of Gene-Targeted Sheep by Nuclear Transfer from Cultured Somatic Cells," *Nature* 405 (29 June 2000): 1066–1069.

13. J. A. Barritt, C. A. Brenner, H. E. Malter, and J. Cohen, "Mitochondria in Human Offspring Derived from Ooplasmic Transplantation," *Human Reproduction* 16 (2001): 513–516.

14. Nell Boyce, "Trial Halted after Gene Shows up in Semen," *Nature* 414 (13 December 2001): 677–678.

15. Eliot Marshall, "Gene Therapy: Panel Reviews Risks of Germ Line Changes," *Science* 294 (14 December 2001): 2268–2269.

16. Mark S. Frankel and Audrey R. Chapman, "Human Inheritable Genetic Modifications: Assessing Scientific, Ethical, Religious, and Policy Issues," American Association for the Advancement of Science, September 2000, <www.aaas.org/spp/dspp/sfrl/germline/main.htm>.

2

The Bioethical Issues

> The real problem is in the hearts and minds of men. It is not a problem of physics but of ethics. It is easier to denature plutonium than to denature evil from the spirit of man.
>
> —Albert Einstein

The term "bioethics" frequently rouses images of academic elites posing as "moral cops," lurking around the laboratories of researchers or in the conference rooms of hospitals, ready to pass judgment on the morality or immorality of practices and procedures. I prefer an explanation that attributes the origin of bioethics to "an increase in the public's interest in a variety of ethical issues that come from medical practices, [which] had given rise to an established 'field' of study in the United States."[1] To a large degree, philosophers, theologians, lawyers, scientists, healthcare professionals, and interested individuals have a hand in the field of bioethics. The ramifications of new advances in genetic medicine are so far-reaching that this interdisciplinary approach, when conducted collegially, serves to inform and facilitate difficult deliberations.

The HGP's July 2000 proclamation that it had completed the map and sequence of the human genome stimulated bioethical questions among laypeople and professionals alike.[2] What it meant was that scientists are now able to identify the location of genes and sequence of bases (A, C, G, and T) comprising the human genome. As mentioned earlier, the immense task of understanding the function of all genes is the work of the postgenome era. This momentous announcement generated feelings of enthusiasm and apprehension. The enthusiasm focused on the promise of advances in treatment for genetic diseases, a positive impact on public health, and improvement in the quality of life in general; the apprehension centered on the concern that the power of DNA was beyond what humans could manage beneficially. Chapter 2 presents the bioethical issues encountered almost daily in the media and in research and clinical practices dealing with genetic disorders. The parameters of purposeful inheritable changes and intergenerational ramifications delineate the matters discussed here.

Many of these issues are admittedly contentious, and there is no way to treat each one adequately in a few pages. I do not raise them to hinder a new technique that promises important therapeutic benefits for those who suffer from otherwise intractable genetic disorders. My intent is quite the contrary. The following chapters discuss the issues with two goals: First, I plan to provide an accurate overview of the ethical implications of this emerging new technology in order to foster understanding. Second, the issues discussed are the foundation for the ethical schema that I develop in this book, the Inheritable Modifications Matrix. The matrix's purpose is to focus public and professional moral deliberation addressing inheritable genetic changes. A summary of the bioethical issues and questions is shown in table 2.1.

GENETIC DISCRIMINATION

Although the stated goal of the Human Genome Project is to develop new diagnostic and therapeutic procedures for genetic disorders, some ripple effects raise concerns. One is the prospect that a diagnosis of genetic disorder could lead to a new form of discrimination or stigmatization based on genes. A recent study assessed the impact of laws restricting health insurance companies' use of genetic information to determine whether these laws decrease the incidence of genetic discrimination and allay the anxiety linked to testing and screening for genetic disorders. The study included a comparative case study analysis, performed in seven states, with different laws regulating insurers' use of genetic information. Results revealed that patients' and clinicians' anxiety about genetic discrimination greatly exceeds reality.[3] Even though laws prohibit discrimination based on genetic information, responses indicate that individuals still worry. One rea-

Table 2.1. Bioethical Issues: How, Where, and Who Will Decide?

Issues	*Questions*
Genetic Discrimination	• Can legislation forestall genetic discrimination? • Is it possible to counteract the "different equals defective" stigma?
Predictive Knowledge	• Should geneticists place a gene into a human for any purpose they deem good? • Can it ever be known with confidence that genetic modification will result in more benefit than harm? • When present humans attempt to predict the characteristics that future generations will value, what standards will justify their choices?
Parental Autonomy	• Is the capability to produce healthy children a right? • Who will determine whose rights take precedence, the parent's or the child's? • What changes are best for the child? • Who decides what standards justify those changes?
Prenatal Testing	• Who and what will preclude the use of prenatal diagnosis from becoming a means to reject undesirable life? • Can parents make decisions based on genetic testing without pressure from insurance companies or society?
Slippery Slope	• Can the downhill trajectory from ethically permissible treatments to unethical alterations be averted? • What regulations and safeguards can curb the potent allure of using genetic information merely for fame and fortune?
Eugenics	• Should the composition of the human genome be left to chance? • Are there moral justifications for positive eugenics? • What safeguards can prevent the reoccurrence of enforced eugenics?
Human Embryo and Stem Cell Research	• Does the human embryo have rights? • Does medical advancement justify ethically questionable research?

son for this is distrust of current legislation until it is tried and upheld in court. At present, there is anecdotal evidence of discrimination by employers and health insurance companies based on genetic testing of individuals, but no cases have yet gone to trial.

A second source of concern involves genetic discrimination in the workplace. As genetic testing becomes more pervasive, employers could require screening of potential employees for the predisposition to genetic conditions triggered by exposure to substances prevalent in the workplace, such as smoke or chemicals. Based on test results, employers could refuse to hire, or require the transfer of, employees carrying genes triggered by such substances, and reasonably claim that it is to the employees' benefit. Science has yet to pinpoint all of the intricacies of interactions between genes and environment in the development of disease, but there are clear relationships established between some specific environmental stimuli and the incidence of disorders in given genotypes. Therefore, employers could make work-related decisions predicated on reasonable medical risk, prior to the manifestation of genetically influenced disease.

A third arena of potential genetic discrimination involves the treatment of persons living with genetic disorders. The concern is that the healthcare insurers and society in general regard persons living with disabilities as less valuable and less worthy of public support than persons without apparent disabilities.[4] The abortion of fetuses considered "disabled" on the basis of prenatal testing sends a disparaging message to persons living with those disorders. The interpretation is that some individuals are too "defective" to exist and are unworthy of the societal resources required to sustain them. Selective abortion of fetuses that carry "defective" genes evokes this concern, especially among the disabled community. Genetic testing and screening programs that result in the termination of prenatal life or the limitation of access to healthcare services will have serious implications for persons presently living with disabilities. This concern is magnified when one imagines that this trend could be extended to include debilitated, mentally incapacitated, elderly, or other marginalized groups in society.

Moreover, skeptics worry that the compilation of a so-called model genome will ultimately define health and disease. There is a history in society and medicine of equating disease with abnormality, of treating abnormal individuals differently, and of showing intolerance toward abnormality by likening it to a substandard state. Geneticists counter the allegation that their intent is to construct an archetype of the perfect genome. Instead, they claim that their purpose is to establish a database prototype of the average genome, which necessarily contains many mutations and defects. There is no "normal" genome since all "healthy" humans carry several abnormal or debilitating genes. Scientists contend that a standard genome will reflect the diversity and variability of the human species, not a model of perfection.

The compilation of an average genome also conjures up worries about the presence of values and social constructions operating in the definitions of normality and abnormality, and of health and disease. The routine association of abnormality with disease imposes a subjective value judgment on a physical or genetic state. An example illustrates the cross-cultural variance in the classification of normality and abnormality. Medical indicators interpreted as health in one country can be categorized as disease in another. American doctors regularly diagnose high blood pressure as an indication of heart disease and low blood pressure as a manifestation of health. In contrast, German doctors regard both high and low blood pressure as conditions calling for medical treatment.[5] This is not to say that the diagnosis of deviation from a norm as disease is inherently subjective. This example merely illustrates that contemporary cultural standards shape beliefs that, in turn, shape perceptions about bodily processes.[6] The presence of social standards, in many cases specific to a culture and a time period, make the definitions of health and disease difficult to pinpoint. As such, current standards of normality and abnormality are a poor basis upon which to categorize existing persons or select the genetic composition of future existents.

PREDICTIVE WISDOM

Hans Jonas is a twentieth-century philosopher who proposed an ethic that emphasizes the role of wisdom in the genetic age. The significance of predictive wisdom, according to Jonas, lies in acknowledging the fallibility of scientific knowledge and the limits of human ability to predict the ramifications of that knowledge. His purpose is to underscore the reality that biotechnology cannot solve all ailments, could potentially increase suffering, and is limited by the boundaries of human prognostication. Jonas contends that it is not possible to know with certainty the future results of present actions; therefore, it is unfeasible to predict with confidence what conditions or social standards will influence the lives of our remote descendants. It is, as Jonas points out, not possible to assess accurately the benefits and risks our genetic decisions hold for future generations, about whom we know almost nothing.[7]

Jonas also emphasizes the importance of acknowledging the not-so-distant limits of human scientific and predictive ability. Medical treatment always involves unavoidable risks and unpredictable outcomes, and likewise genetic alteration of human cells entails unforeseeable consequences. The mechanisms operating in genetic control of growth and development are so intricate that relatively little is understood about the interactions and regulation of a cell's genetic machinery. This is an area of great concern in somatic cell changes; inheritable genetic modification exacerbates this predicament in that it entails both unanticipated and often irreversible mistakes. Animal models demonstrate that genetically altered mice manifest unpredicted side effects, including multiple gene insertions, higher mutation rates, and an increased predisposition to cancer when compared with normally procreated counterparts. Extrapolating these results to humans suggests that genetic alteration will be risk-laden, at least initially. In fact, the foreknowledge of exactly how the human genome responds to intervention may forever be elusive.

As an illustration, let us return briefly to the linguistic metaphor used earlier to explain the structure and function of genes. Classic literary works that withstand the test of time earned that status due to the multiple strata at which they can be interpreted. For example, a short story by Leo Tolstoy may read as a simple tale that a child comprehends, a complex work that requires adult interpretation, or a moral treatise on life. Analogously, there are multiple strata to the human genome that render the interpretation fascinating. Science is probably somewhere between the stage of simply reading the sequence of nucleotide bases and the more difficult task of functional genomics, or understanding what genes do and how they affect health and disease through the production of proteins. The challenge of understanding the genome as a treatise on life, or the level at which genes interact dialogically with other genes and with environmental factors, may forever be extremely difficult. The mandate of predictive wisdom focuses attention on this reality.

This discussion of the unforeseen consequences of inheritable genetic modification does not constitute a charge to procure all relevant predictive knowledge before implementing any genetic technologies. That would inhibit all research and almost all medical procedures. Rather, Nobel laureate Marshall Nirenberg suggested in 1967:

> My guess is that cells will be programmed with synthetic messages within 25 years. . . . The point that deserves special emphasis is that man may be able to program his own cells long before he will be able to assess adequately the long-term consequences of such alterations, long before he will be able to formulate goals, and long before he can resolve the ethical and moral problems which will be raised. When man becomes capable of instructing his own cells, he must refrain from doing so until he has sufficient wisdom to use this knowledge for the benefit of mankind. . . . The decision concerning the application of knowledge must ultimately be made by society.[8]

PARENTAL AUTONOMY

Parental autonomy involves parents making proxy decisions on behalf of their children and distant progeny, plus the issue of the rights of parents to bear healthy children. The right to pursue genetic alteration is considered an expansion of the autonomy possessed presently by individuals in reproductive choices. In most cases, parents have a right to bear and rear children and to make decisions concerning their health, education, and welfare. Does it then follow that parents have the right to pursue forms of inheritable changes that promote health, prevent disease, or enhance traits influenced by genes? If so, everybody is a potential candidate in a system where the market is endless and the financial resources for health care are finite.

Given the possibility that inheritable genetic modification could become a safe and effective procedure within the next two or three decades, parents will want the right to protect offspring from the ravages of genetic disorders. Moreover, the time will come when parents will want to preselect other characteristics of their offspring such as intellect, appearance, or behavior traits. Unfortunately, much of the hype about "designer babies" is built on the common misconception that a child is the result of systematic parental choices of certain genes that influence specific physical or behavioral characteristics. Genes do not always respond in the expected manner, producing a child with traits that meet precise parental standards in the same way that one orders a car or a computer. There is plenty of evidence that genes don't always work in a predictable fashion and that there could be unwanted surprises in store for demanding parents who are willing to pay for only the best in their offspring. Genetic modifications intended to improve a child might go awry, inflicting harm on the child rather than achieving perfection.

Moreover, parents objectify a child when they meddle with their offspring's natural development to accommodate their own expectations. Genetic interventions that are responsible hopes to improve an offspring's future differ from those that are overbearing ambitions to fulfill a parent's personal goals. First and foremost, genetic attempts to improve children genetically must be understood from the perspective of improving the life of *the child* for that child's good, not for the good of the parents or of society. Parents' autonomy in making genetic changes that in some way alter a child's capabilities or life experiences should be tempered by consideration of what the child would realistically want, and should allow for a reasonable level of individual self-determination.

PRENATAL TESTING

The discussion of parental rights leads directly to the issues of prenatal testing and screening. The ability to evaluate an embryo or fetus genetically will increase dramatically as researchers link more and more genes to disease. However, until science overcomes the hurdles of safety and efficacy in genetic alteration, the issue of presymptomatic diagnosis when there is no known treatment haunts this issue. At present, selective abortion is the only available "treatment" for a serious genetic disorder diagnosed prenatally. Leon Kass, professor of ethics and chairman of President Bush's Bioethics Commission, predicts, "Determining who should live or die on the basis of genetic merit is a godlike power already wielded by genetic medicine. This power will only grow."[9] For those opposed to abortion on moral grounds, the termination of a pregnancy for any reason is unacceptable; in addition, many who typically favor abortion rights are uncomfortable with the concept of selective abortion for fetal defects. The distinction seems to involve a different level of culpability, and women who generally accept abortion may feel differently about aborting a particular disabled fetus. Philosophically, it is very difficult to justify selective abortion of a disabled fetus based on a quality of life standard for the child, for the quality of an impaired life is hard to compare to not being born at all. Statistics reflect that only 3–5 percent of prenatal diagnoses reveal a disorder so severe that there is no possibility for a life free from suffering.[10] This means that in many instances, selective abortion is offered as a remedy for moderate disorders or for undesirable characteristics. Whose interests are really served in these cases?

Health insurance companies also misuse genetic information from prenatal screening and testing programs. The case of a Louisiana couple whose first child was born with cystic fibrosis (CF) illustrates this point. The wife became pregnant with her second child, and prenatal testing diagnosed the fetus with CF. Given this information, the family's HMO demanded that they abort and threatened to withdraw coverage from both their first child and the newborn if the couple did not comply with their order. Only after the couple retaliated with the intent to sue did the insurance company retract their threat to discontinue coverage of the family.[11]

At the opposite end of the ethical spectrum, prenatal testing is also a means for parents to produce "better" babies. Prenatal diagnosis is increasingly sought without medical indications of advanced age or a family history of genetic disorder. While in some cases there is pressure from insurance carriers to eliminate disease, in many cases the concern is simply parental anxiety or the desire to choose the sex of the offspring. The practice of sex selection to produce males is widespread in China and India, but remains ethically unacceptable in the United States. However, the trend for families to desire improved babies, also called designer babies, is on the rise. Parents requesting prenatal testing are predominantly middle- or upper-class and well-educated, reflecting a well-documented inequality in the use of these medical services. Some geneticists contend that this trend is acceptable in that the procedures are safe, are personally financed, alleviate anxiety, and uncover some genetic defects that would otherwise go undetected. However, genetic inequities could intensify the issue of discrimination discussed above in that undesirable traits could become a mark of low social class because the bearer's parents had neither the knowledge nor the financial resources to prevent them.

SLIPPERY SLOPE

"Slippery slope" is a term that is often inappropriately used in bioethics to arouse fear, rather than to describe an inevitable trajectory from an ethically questionable act to an unethical one. The misapplication is possibly due to the metaphorical character of the term. Used appropriately, slippery slope represents the concern that the legitimization of some forms of action, such as gene therapy, could lead to other applications that are unethical, such as the production of a "gene rich" class. The implication is that once one crosses the line, it is impossible to prevent unethical practices in genetics. In support of this notion, there is scant historical evidence to illustrate the successful limitation of technology, with the exception of the atomic bomb. To the contrary, most technology advances from research science to development engineering and, finally, to standard commercial products relatively quickly. With time, even controversial technologies, once they are perfected and somewhat affordable, quickly gain acceptance and the initial concerns disappear.

Moreover, the small number of cases where inheritable genetic modification is the only option generates concern about the slippery slope. Infants born with a severe genetic disorder represent a comparatively small percentage of births, and, of those, only two categories of parents are unable to produce a healthy child through somatic cell alterations or preimplantation genetic diagnosis. The first category includes those couples who are homozygous for a genetic disorder that is developmental in nature. This means that both parents carry two copies of a dysfunctional gene and, therefore, have a 100 percent chance of passing the disorder on to their progeny. The second category consists of those couples who object to preimplantation screening and selective abortion on a moral basis.

As a result, there is the concern that the large number of public and private dollars invested to develop this technology target something other than the relatively small numbers of therapeutic interventions required by the couples described above, namely the production of a superior class. Statistics cited by Gail Ross and colleagues serve to mitigate this concern. Their research suggests that relatively rare genetic disorders involving a single gene, such as Huntington's disease, attract only a small percentage of research dollars.[12] Cancers and HIV, disorders that are less directly responsive to gene therapy but that affect a much greater percentage of the population, have greater appeal for commercial funds and National Institutes of Health (NIH) grants. This evidence should alleviate some of the concern generated by inheritable genetic modification, given that there is a justifiable therapeutic application.

The debates swirling around inherited genetic technology range from anxiety about the slippery slope to denial of any reason for concern. Those who act unconcerned believe that adequate safeguards can prevent the downhill trajectory, but those who adhere to the notion of the slippery slope claim that stringent regulations are not sufficient to stem the tide of scientific commercialization. There is yet a third opinion that claims that giving a child enhanced genes is ethical: they contend that genetic enhancement is not fundamentally different from providing a child with an Ivy League education, since both enhance a child's future potential. I find a comparison of education to hard-wiring a child's genes unconvincing. Environmental influences that encourage the development of capabilities that children already possess is decidedly different from genetic interventions that attempt to produce children that are fundamentally different from what they otherwise would have been. Moreover, one need not be a radical egalitarian to object to a society in which the affluent add genetic improvements to the other advantages that come with wealth, while other children whose parents could not afford enhancement fall farther behind. The slippery slope effect witnessed in the eugenics movement in Nazi Germany and the United States in the twentieth century renders this third position still less convincing.

EUGENICS

Eugenics is the planned improvement of the human race through deliberate genetic selection. It is defined simply as good breeding. Eugenics ideology began with Plato's depiction of the ideal state in *The Republic,* in which he asserted that the state is philosophically obligated to regulate marriages to eradicate weak male infants.[13] This practice was actually carried out in Sparta, where weak males were eliminated in an attempt to create the perfect state.

In the twentieth century, eugenics is associated with the Nazi atrocities carried out in the name of racial hygiene during World War II. However, there is a history of American eugenics that is often unknown or dismissed. In 1907, Indiana passed the United States' first sterilization law based on genetic defects. This law permitted the forced sterilization of individuals suffering from undesirable genetic conditions. By the conclusion of World War II, thirty additional states passed comparable legislation that eventually culminated in some 50,000 sterilizations of individuals with objectionable traits or disorders. During this same period, laws founded on eugenic premises barred undesirable immigrants from entrance into the United States and banned interracial marriages in this country. Nazi eugenics differed from American practices in its magnitude and brutality, but philosophically its aims were comparable: to encourage breeding of the genetically fit and to prevent breeding of the genetically unfit in an attempt to cleanse and improve the gene pool. Also, in both countries geneticists and physicians supported and implemented the eugenics programs.

The allegation of eugenics plagues new genetics research, and the publicity results in reasonable concern and unjustified fears. The concern stems from the potential for genetics to prompt new forms of discrimination, and the fear is that scientists will attempt to redesign human beings with enhanced traits or that governments will eliminate undesirable groups of persons. There is probably little basis for worry that humans will direct their own evolution or generate a superior race or class. Traits that comprise the human person, such as personality, intelligence, and appearance, are exceedingly complex. Tens or hundreds of genes act in concert in presently unidentified ways to produce each distinctively human trait. In addition, spontaneous mutations will always bring about unexpected changes in the most perfectly engineered genome. That said, it is impossible to guarantee that individual changes in assisted reproduction will not raise societal standards, or that a government will not mandate the obligation to rectify genetic disorders through the elimination or alteration of undesirable genes. The opposite eugenics scenario is also cited. Leroy Walter, bioethicist and former chair of the Recombinant DNA Advisory Committee, speculates that inheritable genetic modification is not likely to be used to create a super race of geniuses. Instead, he worries that it could be used for malevolent purposes, since "[i]t is easier to destroy or weaken than it is to enhance."[14] Given the heinous history of eugenics in the twentieth century, the ethical dilemmas besetting the enhancement or degradation of humans are difficult to dismiss.

HUMAN EMBRYOS AND STEM CELL RESEARCH

Embryonic stem cells, derived from early embryos or fetuses, are undifferentiated cells that are capable of developing into any tissue of the human body or, theoretically, a complete person. These cells are exciting because they can be steered with various hormones to replace failing cells or regenerate new organs, potentially providing treatment for patients suffering from devastating disorders such as cardiomyopathy, Alzheimer's disease, Parkinson's disease, Type 2 diabetes, osteoporosis, and many others. This topic is actually misplaced in a discussion of inheritable genetic modification, since stem cell research does not entail purposeful genetic alteration. I raise the issue here because the fervor aroused by the potential of embryonic stem cell medicine is reminiscent of the early promise of gene therapy.

Until Jesse Gelsinger died as a direct result of an experimental gene therapy trial at the University of Pennsylvania, many believed that genetic alteration was a panacea for the scourges of humankind. While many tout the tremendous promised benefits of embryonic stem cell research, researchers must overcome serious issues of safety and efficacy before patients realize actual clinical benefits. First, there is much to learn about stem cell biology and the mechanisms that regulate their self-renewal. It is unknown if stem cells will differentiate properly after transplantation, form tumors, or otherwise develop inappropriately *in vivo*.[15] Second, in order to maintain an embryonic stem cell culture in an undifferentiated state, the stem cells must be grown on mouse fibroblast cells.[16] Until science develops another culture medium, the transfer of embryonic stem cells into humans for therapeutic purposes is considered a xenograft by the FDA. Given the infection risks of interspecies transplantation, clinical benefits are far-off and speculative. Predictably, the course of stem cell research will continue to follow that of genetic modification, encountering exciting discoveries and devastating setbacks along the road to new therapies.

Embryonic stem cell research raises ethical concerns because the procurement of these cells entails the destruction of early human embryos. The ethical assessment of embryonic stem cell research fluctuates depending upon the acknowledged status of the human embryo. The most widely agreed-upon viewpoint holds that the human embryo is a potent symbol of human life that deserves profound respect. However, individual interpretations of respect vary extensively. For those who believe that life begins at the moment of fertilization, the derivation of stem cells constitutes abortion and an affront to the dignity of all vulnerable human life; others hold that the embryo is too rudimentary in development to have any moral status at all.

The intent here is not to resolve the question of the moral status of the human embryo, but to point out two biological characteristics upon which there is agreement: the human embryo is alive and human. The embryo is alive since it grows and develops when implanted in a womb. It is human because it possesses a complete complement of forty-six human chromosomes with a unique genetic signature. The human embryo is radically different from a sperm or egg or any other cell in the body because it alone has the inherent teleology to develop into a human being. Biologically, the human embryo is an organism with an independent internal genetic code that directs self-actualization.

While embryonic stem cells are key to understanding the biology of human differentiation and disease, there is mounting evidence that therapeutic applications are likely to come from adult stem cells with none of the moral problems associated with the destruction of human embryos. The range of plasticity demonstrated by adult stem cells was unknown until evidence early in 2001 established that adult stem cells from diverse sources give rise to different cell types to repair the body.[17] Researchers claim to have found what they hail as the "true" stem cell, with unexpected regenerative powers, in adult bone marrow. Bone marrow stem cells can develop into liver, lung, GI tract, and skin cells, contributing to the clinical treatment of disease and tissue repair.[18] Researchers also reported that transplanted bone marrow cells could be reprogrammed to repopulate dead heart tissue in mice.[19] Evidence from adult stem cells derived from fat reported similarly promising findings, for unwanted human fat contains stem cells that can develop into bone and cartilage in the laboratory.[20] Most important, adult stem cells garnered from a patient's own bone marrow or fat tissue avoid the problems of rejection of transplanted tissue from embryonic sources. The challenge lies, as it does with embryonic stem cell lines, in making adult stem cells function effectively *in vivo*. If researchers can overcome this hurdle, adult stem cells promise to be a practical, efficient, and therapeutic option that avoids the ethical controversies and societal divisions associated with embryonic stem cell research.[21] At present, no one really knows which sources of stem cells are best or if any of them are good at all. This is an area of ongoing research.

Harold Shapiro, professor of bioethics and former chair of the Clinton National Bioethics Commission, argues that the present uncertainty and disagreement in the scientific arena are mirrored in the ethical arena.[22] He contends that competing ethical convictions in the United States make agreement difficult, and that we cannot escape conflicting ethical views when the proposals of some segments of society remain unjustifiable or unconvincing to other groups. Moreover, he claims that even those ethical precepts that do find favor will fluctuate with changing societal interests. Other new controversial research, clinical, and public policy issues inevitably will arise from rapid advances in molecular genetics and parallel bioethical considerations. The rapid pace of these developments will generate new perplexities and continuing moral deliberation. It is likely that the interface of ever-changing genetic developments and diverse social, cultural, and religious circumstances will strain our moral propositions and challenge our notions of what it means to be human.

NOTES

1. Tod Chambers, "Centering Bioethics," *Hastings Center Report* 30, no. 1 (January–February 2000): 22–29.
2. See: "The Human Genome," *Nature* 409 (15 January 2001): 745–964; and "The Human Genome," *Science* 291 (17 February 2001): 1145–1434.
3. Mark A. Hall and Stephen S. Rich, "Patients Fear of Genetic Discrimination by Health Insurers: The Impact of Legal Protections," *Genetics in Medicine* 4 (July–August 2000): 214–221.
4. Adrienne Asch, "Distracted by Disability," *Cambridge Quarterly of Healthcare Ethics* 7, no. 1 (Winter 1998): 77–87.
5. Arthur L. Caplan, "If Gene therapy Is the Cure, What Is the Disease?" in *Gene Mapping, Using Law and Ethics as Guides,* ed. George Annas and Sherman Elias (New York: Oxford University Press, 1992): 128–141.
6. David Morris, "How to Speak Postmodern: Medicine, Illness, and Cultural Change," *Hastings Center Report* 30, no. 6 (November–December 2000): 7–16.
7. Hans Jonas, *The Imperative of Responsibility: In Search of an Ethics for the Technological Age* (Chicago: The University of Chicago Press, 1984).
8. John C. Fletcher, "Evolution of the Ethical Debate about Human Gene Therapy," *Human Gene Therapy* 1 (1990): 55–68, especially 57.
9. Leon Kass, "The Moral Meaning of Genetic Technology," *Commentary* 108 (1999): 32–41.
10. Ted Peters, "In Search of the Perfect Child: Genetic Testing and Selective Abortion," *Christian Century,* no. 113 (30 October 1996): 1034–1037.
11. Peters, "In Search."
12. Gail Ross et al., "Gene Therapy in the United States: A Five-Year Report," *Human Gene Therapy* 7 (1996): 1781–1790.

13. Plato, "Republic," in *The Collected Dialogues of Plato,* ed. Edith Hamilton and Huntington Cairns (Princeton: Princeton University Press, 1961), 575–844.

14. Leroy Walters in Craig Donegan, "Gene Therapy's Future," *The Congressional Quarterly Researcher* 5 (8 December 1995): 1089–1112.

15. Tannishtha Reya et al., "Stem Cells, Cancer, and Cancer Stem Cells." *Nature* 414 (1 November 2001): 105–111.

16. P. J. Donovan and J. Gearhart, "The End of the Beginning for Pluripotent Stem Cells," *Nature* 414 (1 November 2001): 92–97.

17. E. H. Kaji and J. M. Leiden, "Gene and Stem Cell Therapies," *Journal of the American Medical Association* 285 (2001): 545–550.

18. D. S. Krause et al., "Multi-Organ, Multi-Lineage Engraftment by a Single Bone Marrow-Derived Stem Cell," *Cell* 105 (May 2001): 369–377.

19. D. Orlic et al., "Bone Marrow Cells Regenerate Infracted Myocardiium," *Nature* 410 (2001): 701–705.

20. P. A. Zuk et al., "Multilineage Cells from Human Adipose Tissue: Implications for Cell-Based Therapies," *Tissue Engineering* 7 (2001): 211–228.

21. C. Almeida-Porada et al., "Adult Stem Cell Plasticity and Methods of Detection," *Reviews in Clinical & Experimental Hematology* 5 (2001): 26–41.

22. Harold T. Shapiro, "Reflection on the Interface of Bioethics, Public Policy, and Science," *Kennedy Institute of Ethics Journal* 9, no. 3 (September 1999): 209–224.

3

The Religious Issues

> Reason and experience both forbid us to expect that national morality can prevail in exclusion of religious principles.
>
> —George Washington

Concerns about inheritable genetic modification often draw upon religious imagery. Some groups consider it distasteful or even irrational to link scientific and sacred concepts for the purpose of ethical analysis. This presents a problem, since it means that the use of language referring to "God," "human nature," or the "sanctity of life" relegates the speaker to the realm of the preposterous, extremely conservative, or narrowly religious. Some consider religious thought to be so isolated from rational thought as to hardly count as thought at all. This is unfortunate for two reasons: 1) religious traditions often bring an important perspective to a discussion that otherwise is lacking, and 2) reference to the sacred renders concerns with religious overtones virtually beyond public discussion. In contrast, spiritual beliefs undergird moral reflection for many persons. In principle, there is no reason that judicious reasoning based on a belief in the sacred cannot contribute to serious moral deliberation.[1]

We now want to consider the religious issues that focus on inheritable genetic modification. The discussion is not an overview of the religions of the world; the theologians cited are all Christian, with an emphasis on the Roman Catholic perspective. But, if that single perspective demonstrates that ethical thinking within a religious framework can relate to secular moral reasoning and the science involved in

Table 3.1. Religious Issues in the Genetic Age

Issues	*Questions*
Playing God	• Are humans stewards or cocreators of Earth? • Does genetic engineering exceed the limits of human competence?
Sanctity of Life	• Does every individual possess unconditional dignity, regardless of his or her imperfections?
Human Nature	• Can genetic modification change behavior or impact the propensity to sin? • Is genetic research a tool or a threat to human freedom?
Thy Kingdom Come	• Is the concept of genetic modification a part of the promise of the kingdom of God, or is it inherently dangerous?
Commodification of Life	• Should DNA have a market value? • Does the relegation of DNA to a market commodity devalue human life?

inheritable genetic modification, other religions need to bring their perspectives to bear as well. An outline of some of the religious issues in the genetic age is shown in table 3.1.

PLAYING GOD: CREATOR OR CREATED?

The Human Genome Project frequently triggers the charge of "playing God." This allegation refers to excessive human control over the genome. Words like "meddling," "tinkering," and "forbidden fruit" reflect the suspicions aroused by genetic research. The rebuttal is that doctors play God every time they treat an illness. All medicine involves subjecting the human body to alterations, and virtually all action involves a change in the natural environment. However, as has been repeatedly noted, biotechnology constitutes a level of control that links power with irrevocability. The hand of humankind now participates in the origination and modification of life forms more directly and efficiently than was formerly possible.

The "playing God" claim explores the fundamental theological issue of humankind's relationship to the divine and the dual components of authority and limitation represented in that relationship. Does humankind possess the wisdom and authority to alter our genetic material, or are we entering a realm relegated to God? This question stimulates renewed interest in the age-old biblical concepts of Creator, creation, and human stewardship. Biblical tradition tells us that God created the world and called it "good."[2] It is a world created for human use, with human beings positioned as stewards over it. Jesus gave clear examples of faithful and unfaithful stewards: Faithful stewards augment the little they are given and, consequently, the owner gives them authority over more; they are the joy of the master. Unfaithful stewards disregard the master's plan and are banished into the outer darkness.[3] The use of the word "steward" implies that humans exist as creatures, even though God, the Creator, gave them dominion over all the works of his hand.

The recent progress in intellectual, scientific, and technological endeavors challenges the steward/creator distinction. Some religious thinkers embrace a doctrine of human coparticipation with God in the endeavor of creation. Arguably, this does not necessarily diminish God's preeminence as divine creator; it merely repositions humans as cocreators with God. Reflections on the issue of cocreators from two prominent theologians, Pope John Paul II and James Gustafson, illustrate differing theological positions on this matter.

The Pope's 1994 address to the Pontifical Academy of Sciences hailed the detailed knowledge of the human genome as progress in human intelligence that inspires admiration and wonder. He stated:

> Scientific progress such as that involving the genome is a credit to human reason, for man is called to be lord of creation, and it honors the Creator, source of all life, who entrusted the human race with stewardship over the world. Discoveries of the complexity of the molecular structure can invite members of the scientific community, and more broadly, all our contemporaries, to wonder about the First Cause, about the One who is the origin of all existence and who has secretly fashioned each one of us.[4]

Clearly, the Pope espouses a traditional biblical definition of Creator that emphasizes the sovereignty and transcendence of God. He uses "lord of creation" with reference to the role of humankind, not unconditionally but in the context of steward. In another address, the Pope referred to humans as the "master of creation" and affirms the human "right-duty of mastering creation and itself in conformity with the finalities willed by the Creator."[5] The terms "lord" and "master" avoid the duality of the term "cocreator." This interpretation endorses human intervention in the world within a relationship of responsibility to a sovereign god, with the caveat, "Such an intervention must consequently respect the fundamental dignity of mankind and the common biological nature which lies at the basis of liberty."[6]

In contrast, prominent contemporary Protestant theologian and ethicist James Gustafson holds the view that humans are "shapers" rather than "conformers" to a static creation:

> Both man's thinking about the nature of ultimate being, and his disposition in life are being altered by the awareness that we are "participants in creativity," rather than the tenders and caretakers of something that has been created.[7]

Gustafson contends that humankind in the genetic age possesses a new awareness of freedom and the power to give direction to human development. However, he worries that these new realizations will not be accompanied by a corresponding quality of moral maturity. Gustafson quotes the renowned Jesuit Teilhard de Chardin to summarize the human status in the genetic revolution: "We have become aware that, in the great game that is being played, we are the players as well as being the cards and the stakes."[8]

Despite obvious dissimilarities in the role assigned to humans in relationship to the Creator, there is a consistent perspective espoused by these two theologians. The common thread is the consensus that human beings possess a mind that is capable of reaching God. Both men envision and endorse that human beings, as bearers of such talent, intervene in the world and become its master. Within this framework, the pejorative use of the term "playing God" becomes operative only if scientists ignore the boundaries of human knowledge and understanding and abnegate the responsibility that stewardship entails.

SANCTITY OF LIFE: IN HIS IMAGE

Sanctity of life signifies the ultimate respect we are willing to accord human life. It expresses an earnestness to treat human life with consideration, to recognize its dignity, and to commit to its preservation. The religious grounding of sanctity of life ensues from the belief that God created human beings in his image and likeness (manifest in intelligence and free will), and as such bestowed upon his creation incomparable dignity. Jesus Christ redeemed humankind from sin and earned for them an immortal destiny. These biblical truths connote the source of the worth conferred upon humankind. John Paul II represents the most conservative Christian interpretation of sanctity of life: "The right of man to life—from the moment of conception until death—is the prime and fundamental right. It is at the root and source of all other rights."[9] The Catholic emphasis on the human condition as a unity of body, spirit, and intellect proceeds from this belief.

The concept of sanctity of life should be an integral part of discussions pertaining to inheritable genetic modification, but legislation in the last half of the twentieth century largely removed theological discourse from the public forum. Thus, sanctity of life functions as a side-constraint due to its religious specificity. This is unfortunate, since this concept is unique in its emphasis upon the unconditional dignity of every individual regardless of his or her imperfections. The notion of sanctity of life endorses scientific research that furthers the well-being of humankind as a unity of body and spirit created in the image of God. It refutes technologies that seek to overemphasize the function of the body or mind or dehumanize persons in any aspect. Leon Kass concludes,

> According to Genesis, God, in His creating, looked at His creatures and saw that they were good: intact, complete, well-working wholes, true to the spoken idea that guided their creation. What standards will guide the genetic engineer?[10]

HUMAN NATURE: "WHAT IS MAN?"

Jesuit theologian Karl Rahner commences his work, *Christian at the Crossroads,* with this question. His response to it is extremely thought provoking: "Man is the question to which there is no answer."[11] Yet human beings inevitably seek the meaning of their humanness. Rahner offers a simple response to the unanswerable question; he defines persons by their everyday experience of life. Moreover, Rahner wonders if we still possess the courage to discuss human nature, especially given the present secularization of thought. The contemporary quest to answer timeless questions has shifted, in large part, from an interest in the spiritual to an objective scientific analysis of human DNA. In contrast, Rahner presents human nature as that which comes into being when God's Word became flesh. Humankind is a mystery, because God, who is also a mystery, chose to become incarnate. Therefore, the mystery of the human being is inseparable from the mystery of Jesus Christ, who is the ultimate expression of humanness.

Rahner goes on to cite three components of human nature: transcendence, freedom, and sin.[12] First, transcendence is the inexplicable capacity for the divine within human beings. It is not experienced as a unique spiritual event, but as concrete everyday encounters with the divine. Second, freedom is the human being's responsibility for his or her actions and choices; it is the God-given ability to decide about oneself. Freedom exists for the self-actualization of the individual in the unique totality of her history, and each individual exercises freedom in her response to the question of belief in God. As such, freedom is the capacity of the human being for the eternal. As a free subject, each person must make the decision to believe or to reject God independently.

The third attribute that Rahner sees as ubiquitous in human nature is sin. How does the reality of sin impact the original supernatural creation intended by God for humankind? Rahner uses an interpretation of the biblical account of the Fall to explain how the reality of sin impacts the original supernatural creation intended by God. On the premise of this belief, he sees original sin as the condition of the beginning of humankind, not a projection of the personal guilt of Adam to us. Its occurrence precedes any personal decision and comprises an analogous state of guilt. As such, sin constitutes the absence of the spirit of God, a deficiency that is the opposite of the situation intended by the Creator. Since this analogous guilt is prior to personal decision, it could be misconstrued as the deprivation of human freedom. Instead, Rahner believes that the human being is actualized as a free subject in situations that are always determined by history and other persons.

> But this means that he always and inevitably exercises his personal, inalienable and unique acts of freedom in a situation which he finds prior to himself, which is imposed on him, and which is ultimately the presupposition of his freedom.[13]

As such, individuals are endowed with freedom and limited by necessity.

Rahner concludes that many human characteristics once thought to be the unchangeable structures of human nature are culturally determined and, therefore, mutable. He maintains that human beings must always consider past history plus the future in the quest to determine what humankind is. However, the transcendence and dignity of human beings, for Rahner, are ontological and immutable characteristics that establish for humanity an inviolate sphere of rights and expectations that define humanity's relationship to God, to society, and to the church.

THY KINGDOM COME: THE KINGDOM OF GOD ON EARTH

This concept implies a kingdom of "saints," where God will be the king by virtue of his royal rights and the love of his people. It takes the form of a purely spiritual movement, not of a victorious national uprising as was frequently expected. It began modestly among a small band of believers following the death and resurrection of Jesus Christ and repeatedly arouses opposition and persecution to this day. The Church, as the body of Christ, fosters the development of the kingdom of God on earth. The final eschatological realization of the kingdom will occur when the "saints" are with God in the joy of the heavenly banquet.

The kingdom of God symbolizes a realm of freedom and joy that is devoid of the ravages of tears and pain.[14] The promise is the experience of goodness and justice where God dwells among mortals in the human heart. It is on the basis of this promise that Christians envision a better world. Genetic medicine has the potential to further God's kingdom on earth through the alleviation of tears and pain caused by intractable genetic disorders. If genetic medicine fulfills its near-utopian promise, it will truly initiate a new standard for healthcare and a novel expectation of human existence.

As we consider the future benefits promised by genetic medicine, we must not lose sight of those who may not benefit. Genetic medicine possesses the capability to disenfranchise those individuals without access to the promised benefits or those living with genetic disorders. The utopian search for the perfect child or for a world free from all genetic disorders has the potential to further marginalize the poor and disabled. It will be a grave injustice if advances in genetic medicine, which have the potential to further the kingdom of God on earth, instead result in the creation of new forms of intolerance and injustice.

COMMODIFICATION OF LIFE: MANUFACTURED OR CREATED BY GOD?

Concerns about the commodification of human life stem from attempts to use human DNA merely as a vehicle of production and profit. In the extreme, this practice equates to a "new kind of gold rush" in which "the territory is the human body."[15] The relegation of DNA to market standards is objectionable, since many religious traditions regard the human genome as a sacred entity that is profaned when relegated to a commercial commodity. The pricey manufacture of designer babies, discussed in chapter 2, is a notorious example of the production of human life with made-to-order characteristics.

The patenting of DNA sequences is another example of commodification of genetic material.[16] Until the advent of the genetic age, patents applied only to novel inventions that had an obvious utility. Recently, genes and DNA sequences have been patented despite unidentified utility and controversy over whether they belong in the category of discoveries rather than inventions. In 1980, the U.S. Supreme Court set precedent for the practice of patenting genes in *Diamond v. Chakrabarty* by granting a patent for the invention of a new bacterium with the ability to clean up oil spills.[17] More recent criteria issued by the U.S. Patent Office advocate a limited interpretation of the utility prerequisite, calling for the disclosure of some documented benefit of the patented sequence. Favorable outcomes of gene patenting include incentives for investment and research that result in the production of new drugs and therapies. Detrimental effects include the restriction of access to research by other scientists and licensing agreements that limit clinical use of genetic tests based on patented genes.

Religious thinkers generally do not oppose biotechnology or deny the necessity for financial returns. Their anxiety arises more from a clash of worldviews than from a direct objection to science. Judeo-Christian beliefs clearly distinguish between objects that are bought and sold at a given price in the marketplace and human entities that are not for sale. Judeo-Christian beliefs have historically condemned the practice of selling human organs, human beings, or in any way reducing life to a monetary value. Many fear that the breach of this taboo, in combination with the powerful economic interests operative in genetic technology, will ultimately lead to a literal commodification of life.

In contrast, the biotechnology industry extols its goals as verifiably moral. The industry contends that the incentive for funding enhances competition and expedites research in a way that has not previously been possible. The benefits garnered include augmented profits for the genetics industry and the promise of new cures and designer drugs for life-threatening diseases. Biotechnologists cite these admirable ends as exemplary of the moral status of genetic medicine. The goals are admirable, for human economic and physical well-being are easily justified. Religiously based critiques express concern, as explained above, that the commodification of life may impact other important values not recognized by the market. Specifically, these are values that revere life as more than a chemical sequence and well-being as more than prosperity.

PAPAL WRITINGS: MORAL BOUNDARY OF GENETIC MODIFICATION

The moral direction of John Paul II expresses the reality of an order in the created universe deserving of respect. He invites humans to participate in the work of the Creator, as "the researcher follows God's design. God willed man to be the king of creation."[18] However, the Pope requires that genetic modification of human life respect the order demonstrated by nature and the essence of humankind as created by God. He cited three conditions that must be met in order for human genetic modification to be morally acceptable. These three requisites delineate the limits beyond which genetic modification of the human genome is proscribed. First, the biological integrity of every human being as a unity composed of body and soul must be respected. Second, embryonic life must be accorded the basic rights due all humans. Third, manipulation of the genome may not aim at the creation of new or different groups of people.

The Pope underscores the unity of the human being in order to emphasize the dignity that is due a creature made in the "image of God." His concern is that genetic alterations could result in differences that "provoke fresh marginalization" in our world by enhancing or degrading human traits that compromise the integrity of humans. For the Pope, the purpose underlying the genetic changes is of primary importance.

> The fundamental attitudes inspiring the intervention we refer to should not derive from a racist, materialist mentality aimed at human happiness which is really reductive. Man's dignity transcends his biological condition.[19]

That which is transcendent in the human being, dignity and freedom, must be protected from technological assault.

The Pope's comments reveal a foreboding engendered by the reductionist potential of the new genetics: "The ability to establish the genetic map should not lead to reducing the subject to his genetic inheritance."[20] The treatment of genetic illness must not include only the technical problems posed by the treatment, but also give primary consideration to the "patient in all his dimensions." The term "dimensions" refers to the unity of the human person, the wholeness of the "affective, intellectual, and spiritual functions." In his definition of dimensions, John Paul II includes personal relationships, circumstances, and history.

In further specifying the parameters of the moral limits of human genetic modification, the Pope takes a position on the therapy/enhancement distinction. He endorses therapeutic interventions as long as the harmful effects do not exceed the anticipated benefit:

> A strictly therapeutic intervention, having the objective of healing various maladies—such as those which stem from chromosomic deficiencies—will be considered in principle as desirable, provided that it tends to promotion of man's personal well-being, without harming his integrity or worsening his life conditions.[21]

He does not condemn genetic enhancements that are not strictly therapeutic but "aimed at improving the human biological condition." He requires that two conditions be fulfilled, given the profound moral significance of improving traits contrasted with correcting a defect. First, the intervention must not impair the origin of human life, and second, such an intervention must respect the dignity of humankind and the "common biological nature" that forms the basis of human liberty.

In a more recent address, the somatic cell–germ line distinction took precedence. Adhering to the same parameters of the "transcendental vocation" and the "incomparable dignity" of the human persons that constantly ground his principles, the Pope endorsed medical treatments that ameliorate "fatal hereditary pathologies." In doing so, he endorsed the possible elimination of those pathologies: "By acting on the subject's unhealthy genes, it will also be possible to prevent the recurrence of genetic diseases and their transmission."[22] This statement indicates that inheritable genetic modification aimed at the prevention or transmission of genetic disease and not interfering with the origin of human life could be considered.

Jesus Christ is the incarnation of the moral law. As such, he will always be the norm for embodied human life. Colossians 1:15 states that "He is the image of the invisible God, the firstborn over all creation." Thus, Jesus is a model for moral action and for human physiological life. This is not to imply a normative gender, class, race, appearance, or genotype, but rather a standard for the crucial cognitive and biological elements that are influenced by genes and shared by all human beings, regardless of condition. Ultimately, the unity of qualities that Jesus exemplified must define the moral limits for inheritable genetic modification.

NOTES

1. See Gregory E. Kaebnick, "On the Sanctity of Nature," *Hastings Center Report* 30, no. 5 (September–October 2000): 16–23.
2. Genesis 1:31.
3. Matthew 25:19–30.
4. Pope John Paul II, "The Human Person—Beginning and End of Scientific Research," Address of Pope John Paul II to the Pontifical Academy of Sciences (October 28, 1994), in *The Pope Speaks* 40 (March–April 1995): 80–84.
5. Albert S. Moraczewski, "The Church and the Restructuring of Humans," in *Medicine Unbound,* ed. Robert H. Blank and Andrea R. Bonnicksen (New York: Columbia University Press, 1994), 40–60.
6. Pope John Paul II, "The Ethics of Genetic Manipulation," *Origins* 13 (17 November 1983): 386–389.
7. James M. Gustafson, "Theology Confronts Technology and the Life Sciences," in *On Moral Medicine,* ed. Stephen E. Lammers and Allen Verhey (Grand Rapids, Mich.: Eerdmans, 1987), 35–40.
8. James M. Gustafson, "Christian Humanism and the Human Mind," in *On Moral Medicine,* ed. Stephen E. Lammers and Allen Verhey (Grand Rapids, Mich.: Eerdmans, 1987), 537–582.
9. Pope John Paul II, "The Ethics."
10. Leon Kass, "The Moral Meaning of Genetic Technology," *Commentary* 108 (1999): 32–41.
11. Karl Rahner, *Christian at the Crossroads,* trans. V. Green (New York: The Seabury Press, 1975).
12. Karl Rahner, *Foundations of Christian Faith,* trans. W. D. Dych (New York: The Seabury Press, 1978).
13. Rahner, *Foundations.*
14. Revelation 21:4.
15. Suzanne Holland, "Contested Commodities at Both Ends of Life: Buying and Selling Gametes, Embryos, and Body Tissues," *Kennedy Institute of Ethics Journal* 11, no. 3 (September 2001): 263–284.
16. L. S. Cahill, "Genetics, Commodification, and Social Justice in the Globalization Era," *Kennedy Institute of Ethics Journal* 11, no. 3 (September 2001): 221–238; and Susan Carter Poland, "Genes, Patents, and

Bioethics, Will History Repeat Itself?" *Kennedy Institute of Ethics Journal* 10, no. 3 (September 2000): 265–281.

17. J. Robert Nelson, On the New Frontiers of Genetics and Religion (Grand Rapids, Mich.: William B. Eerdmans Publishing Company, 1994), 68.

18. Pope John Paul II, "The Ethics."

19. Pope John Paul II, "The Ethics."

20. Pope John Paul II, "The Human Person."

21. Pope John Paul II, "The Ethics."

22. Pope John Paul II, "The Human Person."

4

The Philosophical Issues

> What we can do is to make life a little less terrible and a little less unjust in every generation. A good deal can be achieved this way.
>
> —Karl Popper, *Utopia and Violence*

Analytic philosopher R. M. Hare questions whether philosophy is even part of the discussion of current issues such as inheritable genetic modification.[1] He wonders how a discipline that is taught only in academic institutions could possibly apply to concerns of life. In many aspects, his question is extremely appropriate. However, with the onslaught of new discoveries and opportunities in molecular genetics developing at breakneck speed, reflective questioning serves to inform and enrich the continuing debates. Philosophy brings to the table a contemplative generality and a mode of rational inquiry that offers a different perspective from that of science. The questions generated by the juxtaposition of philosophy and inheritable genetic modification are outlined in table 4.1.

GENETIC DETERMINISM

Genetic determinism attributes human characteristics and behavior predominantly to genetics, rather than recognizing that genes are usually only contributing causes. This thinking employs the "blueprint for life" as the main metaphor for the human body, emphasizing the production of specified proteins according to a preordained design. Utilizing this

Table 4.1. Philosophical Issues in Inheritable Genetic Modification

Issue	*Questions*
Genetic Determinism	• Are genetic traits immutable? • Is it *all in the genes?*
Justice	• What does justice mean in the postgenome age? • Can we distribute the benefits and risks of genetic medicine fairly?
Diversity	• Whose and what standards will substantiate the decision to expand or contract genetic diversity? • What measures should protect diversity within the human species?
Genetic Patrimony	• Do humans have a right to an unaltered genome?
Obligation to Future Generations	• Who are the subjects of our ethical domain? • What is our obligation to future generations? • Are there moral considerations that override the rights of existing persons?
Personhood	• Is a *person* different from a *human?* • Are the characteristics of dignity and freedom dependent on the level of human development?

analogy, DNA equates to a computer program that stores and transmits information upon demand. The imagery of genetic blueprints and cellular machinery propagates a new way of thinking that tends to reduce the mystery of human life to a DNA sequence. James Watson, who along with Francis Crick described the double helical structure of DNA in 1957, fostered this viewpoint with statements describing the HGP as the search for "ultimate answers to the chemical underpinnings of human existence."[2] His statement represents the conviction that since scientists cracked the genetic code, the explanation of all phenotypic characteristics (both human health and behavior) will follow.

This notion also originates from the "gene-mania" of the scientific community and the public media when they mistakenly give the impression that there is a genetic solution to all the problems of life. Genetic determinism exaggerates the role that heredity plays in human existence and minimizes the roles of nurture and the human spirit. If human action ensues from a DNA sequence rather than from free will, there are serious implications for education, employment, health care, military service, and criminal law. On the individual level, genetic determinism relegates a person's future morbidity and mortality to DNA. On the societal level, genetic determinism dictates how an individual can live and work in society. This means that important life choices expand or shrink in deference to a genetic verdict.

Some philosophers link genetic determinism to a more sophisticated version of genetic discrimination. This notion stems from the perceived social power of genetic information to perpetuate a new lower class through the creation of categories based on the presence or absence of given genes. No one will deny the possibility that this could happen. However, to limit the discussion to genetics' potential to increase discrimination is too narrow. It may be more helpful to discuss the issue of genetic determinism in terms of the philosophical language of object and subject. Two very dissimilar philosophers, Immanuel Kant and Pope John Paul II, denounce the objectification of persons as objects of research rather than as subjects with unmistakable dignity.

Immanuel Kant formulated the most influential secular principle establishing the human being as a sovereign subject toward the end of the eighteenth century. He mandated that all rational beings should be treated as ends in themselves, and not merely as means.[3] Since rational beings possess an unconditional and absolute value for Kant, he deemed it immoral to use persons simply as a means to an end whose value is only relative. Pope John Paul II echoed the mandate prohibiting the reduction of the human being to an object of technology. He called attention to the reductionist potential of the new genetics: "The ability to establish the genetic map should not lead to reducing the subject to his genetic inheritance."[4] He asserted that science alone cannot account for the "ultimate purpose" of human existence, which is a relationship with God. The Pope claimed that a strictly scientific explanation negates human freedom and clashes with the "irrefutable evidence" that our inner self cannot be reduced to an object, but consistently remains the author of human actions and beliefs.

Thomas Murray, a bioethicist well versed in genetics, likened the reductionist viewpoint to equating a wonderful musical piece to a sequence of black symbols on white paper.[5] Any musician or music lover knows that the beauty and reality of a piece of music lies in its performance, not solely the composition of its notes. The same is true of the human being. Each individual is a combination of genes, some flawed and some functional, acting in concert with environmental factors to direct the development and maintenance of life. Music loses none of its grandeur through its portrayal as a series of notes, yet its true glory is revealed in its performance. In comparison, the decoding of the genome need not reduce human significance to the endless repetition of A, G, C, and Ts. Rather, the expression of the DNA sequence in a unique individual deepens our appreciation for the inexplicable beauty and mystery of human life. The simplistic quality of genetic determinism is especially apparent if one considers that the human genome differs from the chimpanzee genome in less than 1 percent of its genes, yet that small difference in DNA makes a huge difference in the end result.

JUSTICE

The classical definition of justice, "to each what is due," is subject to many different interpretations in the genetic age. By understanding the different philosophical bases from which one approaches considerations of justice, one can appreciate the dissimilar positions on heritable genetic modification that ensue. Let us compare two differing perspectives on the elusive principle of justice as proposed by John Rawls and David Hollenbach, respectively, to illustrate the difference.

The Rawlsian contractarian conception of justice prescribes a fundamental equality of individuals attained by a corresponding equality of opportunity and resources. As such, justice requires the overall equal distribution, or redistribution, of resources to compensate for inequalities in social or physical attributes that are undeserved. Rawls offers a process for determining fair distributions that involves taking the original position behind a "veil of ignorance" where those with decision-making capacity are unaware of their rank in society.[6] The "veil of ignorance" is Rawls' philosophical concept that calls decision-makers to set aside their biases and reflect on what is fair for the broader community from an objective standpoint. Another way of describing this concept is the legal idea of blind justice, which refers to "liberty and justice for all" regardless of race, gender, religion, genetic makeup, or class.

The Greek play *Oedipus the King*, written by Sophocles in approximately 450 B.C., illustrates the concept of blind justice. In the play, young Oedipus, a citizen of Corinth, fled his home in order to evade a frightening prophecy that he would kill his father and marry his mother. During his flight, he attacked and killed a small band of travelers who refused to let him cross at a "place where three roads meet."[7] He eventually completed his journey and arrived in Thebes, a kingdom in need of a leader to free it from the tyranny of the

Sphinx. Oedipus ultimately became the leader of the kingdom by solving the riddle of the Sphinx. He ascended the throne and married the widowed queen of Thebes, Jocasta. Oedipus became known as a just king until his reign was hampered by a devastating outbreak of the plague, a scourge whose only remedy was justice for the murdered king of Thebes, Laius. In pursuit of justice and an end to the plague, Oedipus ordered an investigation into the murder and decreed "upon the murderer I invoke this curse—whether he is one man and all unknown, or one of many—may he wear out his life in misery to miserable doom!"[8] The investigation concluded with the discovery that Oedipus himself was the man who had murdered Laius (his real father) and married Jocasta (his real mother). When his true parentage and crimes were revealed, Oedipus blinded himself in despair and asked for banishment as "the greatly miserable, the most accursed . . . above all men on earth."[9] This ancient play dramatizes Rawls's claim that true justice is blind to social standing or special interests.

While Rawls does not comment about genetic choices specifically, he advocates managing natural inequalities through appropriate social policies and practices. Rawls maintains that natural physical attributes, such as appearance or intelligence, provide enhanced life prospects for some, resulting in inherently lower socioeconomic life prospects for others who do not possess the same advantages. For Rawls, this inequality is unjust because it contradicts the norm of equality of opportunity.

Rawls does not contend that the norm of equal opportunity mandates genetic modification to achieve that goal. His theory tolerates a broad range of inequalities in society and rejects the requirement to ensure the equal welfare of all persons.[10] Likewise, it is fair to say that Rawls would advocate genetic modification to eliminate deleterious genes as a procedure required by justice. Devastating genetic disorders preclude individuals from effective participation in many aspects of society, and no remuneration can effectively compensate for those disparities. According to the principle of justice, social policy is truly just only when it improves the plight of the poorest and most vulnerable, and victims of serious genetic disorders are justifiably among those individuals.

Following this line of reasoning, inheritable genetic therapy could be preferable to somatic cell therapy for two reasons: it offers a cure rather than a temporary treatment, and it removes deleterious genes from all succeeding generations. Yet, even if science resolves the problems of safety and efficacy that currently plague this technique, the consideration of justice tempers the enthusiasm. Inheritable genetic therapy is not an alternative for those persons who are genetically the least well off, those already living with disorders that occurred during development. Their options include somatic cell therapy or other temporary treatments. This presents a moral dilemma, since inheritable genetic therapy benefits only the as yet unborn and those whose parents can or will pay the price. This predicament exemplifies the quandaries surrounding the notion of intergenerational justice and resource allocation.

In comparison to Rawls, David Hollenbach defines justice based on the Christian principles of dignity and mutuality. He delineates three categories of justice. First, commutative justice addresses the area of private interactions among persons. It expresses the need for mutuality based on dignity and equality in personal relationships and agreements. The other two types of justice—distributive and social justice—relate to the public domain. Distributive justice asserts that all persons have some claim to those goods that are public or social, since all members of society join in the creation of these goods through their membership in the human community. Social justice affirms the responsibility of society, through the state, to assure that all persons share in the common good.[11] Applied to inheritable genetic modification, commutative justice permits, but never requires, genetic changes that reinforce human dignity and mutuality in relationships. Distributive justice could endorse only those procedures that are attainable in some share by all, negating the necessity of wealth as a prerequisite for access, and social justice requires institutional standards to guarantee that these norms are achieved.

It is readily apparent that although Rawls and Hollenbach espouse distinct theories of justice, one based in equality and the other in dignity, their views culminate in similar positions with regard to inheritable genetic modification. Both concur that although this technology does not violate the principle of justice in itself, justice is transgressed if a product of society, such as medical treatment, is limited to any individual or class based on wealth or influence. These different interpretations of justice diverge when evaluating enhancement modifications of inherited traits. While Rawls' theory commends enhancement for the purposes of equality, Hollenbach challenges enhancement procedures on the basis of distributive justice and human dignity.

DIVERSITY

Encoded in every species' genome is a stockpile of genetic options that ensure the survival of successive generations under varying conditions. In other words, genetic diversity enhances the success of a species. When artificial selection systematically eliminates genes that appear useless or undesirable in the present, the effect can result in the unwitting destruction of options that permit species survival into the unknown future. The attempt to eliminate the gene for sickle cell anemia from individuals of African ancestry illustrates this phenomenon. While sickle cell anemia is a devastating disease for those who have two copies of the gene, carriers who have only one copy of the gene have an increased resistance to malaria and no symptoms of sickness. Thus, if it were possible to eradicate the sickle cell gene from the population, it would entail a life-threatening change for those who live in regions of the world that are susceptible to malaria. Admittedly, genetic alterations of this type will never eliminate any given gene completely, given the dual restraints of limited resources and spontaneous mutations. Yet, the impact of decreasing variation through the elimination of undesirable genes or deleterious genes often has a downside that is as yet undetermined.

The manipulated evolution of corn is another example that illustrates the benefits of genetic diversity and the harm of uniformity. Colonial and modern corn production over the past 500 years has produced more abundant harvests, but also a decline in the genetic options encoded in the corn genome. Human-driven evolution of agricultural products operates through the artificial selection of desirable traits and the elimination of deleterious ones. Its goal is to enhance productivity, but its overall effect is to reduce the variability within the species. This is not necessarily problematic in agriculture as long as the environment and human preferences remain within a given spectrum. However, the reduction of diversity can be onerous from an evolutionary standpoint, since genetic variability is a crucial component of long-term survival. Genetic diversity, in both human and nonhuman species, is a valuable planetary resource that is in our best interests to safeguard and preserve.

Hans Jonas echoes this conviction in his philosophical approach to biotechnology. For Jonas, genetic diversity is "biodiversity" that encompasses "humans within nature." He mandates a preservation of biodiversity for all aspects of the world, including genomic, species, ecological, political, economic, and cultural diversity. He believes that biodiversity is crucial from three standpoints: From an ecological and evolutionary perspective, it is important to "hedge our bets" for a long-term adaptive advantage. Ethically, it is important to protect diverse "values, obligations, and responsibilities." And instrumentally, biodiversity is valuable as an "economic, agricultural, healthcare, and recreational resource."[12] Human activity is rich, complex, and intimately intertwined with nature. On every level, existence depends upon the dynamic interplay of the natural world and human beings. The protection of biodiversity is essential to that dynamic interdependence.

GENETIC PATRIMONY

Patrimony is a concept that is more familiar to Europeans than to Americans. Our word "heritage" is very similar to, but not nearly as complex as, the French word *patrimoine,* which indicates the totality of a person's assets bequeathed to his or her heirs. This word sometimes relates to genetic endowment, given its reference to inheritance and property. *Patrimoine génétique* (genetic patrimony) is, in a figurative sense, synonymous with genetic material because it signifies the collective context, the human genome, as contrasted with individual legacy, the gene. Opponents of inheritable genetic modification use the notions of heritage and ownership, as implied by *patrimoine génétique,* to emphasize the basic human right to an unmodified genetic endowment. While an autonomous subject could conceivably manipulate his or her own genetic endowment, for example through somatic cell changes, the authorization of inheritable genetic modification seems to tamper with an inalienable right at the collective level. Genetic patrimony argues that the unique worth of an unmodified collective genetic heritage ranks above individual interest.

Let us consider an analogy to further elucidate this concept. When the expression *patrimoine culturel de l'humanité* is used, it symbolizes European cathedrals, Greek temples, classical literature, and magnificent works of art that belong to the ages. Such assets of humankind are irreplaceable, immeasurably valuable, and warrant specific forms of social protection that surpass individual interest. Therefore, "Just as an individual cannot tear down a Gothic chapel for his private convenience, one cannot touch the 'genetic patrimony,' even if some persons would benefit."[13] No one questions the amelioration of devastating genetic disorders for the sake of preservation of life; analogously, there is no argument opposing the repair or restoration of a beautiful cathedral that would cease to exist if neglected. Rather, *patrimoine génétique* expresses the intrinsic value of the irreplaceable heritage of humankind and the importance of preserving it for future generations. The sense of this term does not connote that this heritage must be unsullied or flawless; it is an expression of the value of protecting our common genetic legacy from harmful manipulation.

OBLIGATION TO FUTURE GENERATIONS

The medical profession is morally obliged to employ the most effective methods and technology available to alleviate or prevent genetic disorder. With the advent of genetic medicine, there is pressure to extend this obligation to the health of forthcoming generations. The potential power of direct genetic intervention brings the question of this obligation to the forefront in discussions of normative ethics. These discussions generally use the language of rights of future individuals, debate whether those rights can ever trump obligations to existing persons and their immediate progeny, and often degenerate into a struggle among incommensurable duties.

The problem lies in the limitations of normative ethics, deontological or utilitarian, to resolve the conflict. Deontology is an ethical theory that emphasizes what we are supposed to do; it posits that some actions are right or wrong in themselves and not simply because of their consequences. Deontological ethics, based in the obligation or duty of the ethical agent, holds that certain actions are intrinsically good or bad regardless of context or consequences. Duty privileges autonomous individuals, but regards persons as "ends-in-themselves" only when they exist. Therefore, duty is difficult to extend to future generations. Utilitarian ethics focuses on the maximization of average utility or happiness. This is difficult to extrapolate to future generations, since it relies on an established number of individuals and an assessment of future happiness that cannot be presently known. Another way to calculate utility is to estimate total, rather than average, happiness. This means of calculation is attractive since it bypasses the question of unknowable numbers, but it presents additional problems.

Total utility offers no guidance as to when obligations to future generations take precedence over obligations to present existents or what constitutes happiness in the distant future. At present, there are no agreed-upon moral resources to address these biotechnological challenges.

Contemporary philosopher and ethicist Jere Surber makes a substantial contribution to the consideration of obligations to future persons. His recommendations do not solve all the problems discussed above, but serve to demonstrate the important philosophical issues surrounding this topic. First, Surber suggests the phrase "responsibilities to future generations (or individuals)" as a place to begin;[14] "responsibilities to future generations" corroborates the ethic of responsibility advanced by Jonas without the requisite of obligation or moral duty. Second, Surber posits that an analysis of this dilemma differentiates between "affirmative" and "prohibitive" actions as a means of clarification. An "affirmative" action is one that preserves some good for our progeny; "prohibitive" actions are those with a long-lasting negative impact. Given the difficulties with determining what constitutes an "affirmative" action for future societies, Surber suggests that clearly "prohibitive" actions, such as the pollution of the world's water supply, are more suitable to identify responsibilities to future generations. Third, and especially pertinent, Surber questions whether or not there are identifiable "constant factors" that provide a basis for transgenerational obligations: Are there identifiable needs or values that comprise the parameters of responsibility to future persons? Finally, Surber cites the issue of predictability discussed in chapter 2 under predictive wisdom. Predictability refers to the possibility of ever reliably knowing the impact of one's actions, especially when considering future unknowns.

Surber's final suggestion seems to summarize the intricacy of intergenerational ethics appropriately. His comments underscore the complexity of the repercussions of present actions upon the freedom of ensuing generations:

> It would be an error, I think, to impose our own "social ideal" upon future generations, but it would be equally wrong to avoid taking any action at all on the grounds that it would tend to abridge their own freedom of choice. Rather, it is always crucial to consider what sort of balance could be achieved between positively enhancing the quality of life of future generations and leaving open possibilities for their own determination.[15]

Many actions influence affirmative or prohibitive possibilities for future generations, often without appreciating the difference. Surber designates the balance between the two as the duty to ensure that future generations are not "worse off than we are at present."[16]

PERSONHOOD

The concept of personhood focuses on the distinctive properties, metaphysical and moral, that confer moral status on an individual. Metaphysical personhood consists of psychological properties such as self-consciousness, speech, sensations of pain and pleasure, emotion, and free will. Moral personhood adds the requirements of the capacity to make judgments of right and wrong and display moral motives. The belief is that some special combination of properties confers a moral status upon persons that is different from humans or nonhuman animals.

Philosopher and bioethicist Tristram Engelhardt makes the distinction between persons and humans by identifying a line between biological human life and personal human life. According to Engelhardt, "human" is a biological distinction that indicates membership in the species *homo sapiens,* but "persons" are those individuals who are rational, self-determining entities. Engelhardt is explicit on the implications of personhood. Persons are significant, and humans are not, because persons constitute the moral community. Engelhardt is careful to define the standards necessary for beings to have membership in this community: members are "entities that are self-conscious, rational, free to choose, and in possession of a sense of moral concern."[17] By definition, it is only persons who are capable of understanding when they or others act morally or immorally. Consequently, for Engelhardt, only persons capable of moral discussion are subjects who have rights and must be treated as ends-in-themselves. In this context, to be treated as a subject one must be a person, not merely human. The concept of personhood relegates the human embryo and fetus to a minimal moral claim based on their primitive stages of development. The "minimal" moral claim is likened to that elicited by an animal in a comparable stage of development.[18] This position stands in direct opposition to proponents of the sanctity of life viewpoint, who hold that the human being, from conception to death, cannot be exploited for any purpose whatsoever.[19]

The personhood discussion represents the attempt to equate being *with* consciousness, as contrasted with the differentiation of being *from* consciousness. John Crosby, Catholic philosopher, embraces the differentiation of being *from* consciousness position to substantiate the sanctity of life viewpoint. Crosby refers to the discernment of Socrates in the *Theatetus* to illustrate the distinction. Protagoras stated: "Man is the measure of all things—alike of the being of things that are and of the not-being of things that are not."[20] From there, Protagoras proposed that things are to each person as they are perceived. Socrates took a contrasting viewpoint by demonstrating that persons can err in their perception of objects. To illustrate, he explained the case of a man who perceived the wine to be sweet when he was well and found the same wine to taste sour when he was sick. His point was that there are objects that are not measured merely by our discernment of them, demonstrating that it is possible to distinguish between objects and our current perception of them.

Crosby parallels this argument to demonstrate the distinction between the being of a person and the consciousness of a person. He argues that just as an object cannot be reduced merely to our perception of it, the human person resists reduction to "conscious self-presence." Crosby utilizes this argument to distinguish between a being as a person and a being consciously present to itself. He uses the examples of sleep and temporary coma to illustrate this distinction.[21] When a person regains consciousness from sleep or temporary coma, the experience is not one of regained personhood but of recovered consciousness. He explains:

> The conscious life of the person is not the whole person; it is that in which the being of the person is actualized, and this implies that a person is more than consciousness, and that his or her being is to be distinguished from his consciousness.[22]

Crosby's argument refutes the limitations of the concept of personhood in an attempt to validate the status of the human embryo. As such, the human embryo that lacks consciousness is a person, differing only in the stage of development.

The differences in the interpretations of personhood by Engelhardt and Crosby are readily apparent. These distinctly contrasting methods of assessing moral worth are at the core of current philosophical debates concerning the morality of abortion, euthanasia, assisted reproduction, and experimentation of human embryos.

These different approaches to the concept of personhood illustrate the way in which thoughtful persons see morality differently, resulting in seemingly interminable conflicts in public discussions and policy-making attempts.

This review of the ethical issues raised by inherited genetic modification from the bioethical, religious, and philosophical perspectives demonstrates areas of optimism and concern. The promise of new and more effective diagnoses and treatments for many common disorders generates tremendous optimism. On the other hand, the shared concern is that the irresistible temptation to know and control through genetics may challenge our notion of what it means to be human. Bioethicists, theologians, and philosophers alike are asking: How can we safeguard what is uniquely human, recognizing that manipulation of the human genome is inevitable? How can we draw the line? Since it is infeasible to define humanness, it is ethically beneficial to identify other means to delineate propitious boundaries to safeguard humankind. The second part of this book responds to this requisite by exploring what type of people we are, or ought to be, through the pursuit of an ethical perspective steeped in virtue and inclusive of duty, particularity, and utilitarianism.

NOTES

1. R. M. Hare, "Ethical Theory and Utilitarianism," in *Utilitarianism and Beyond,* ed. Amartya Sen and Bernard Williams (New York: Cambridge University Press, 1982).

2. James Watson, "The Human Genome Project: Past, Present, and Future," *Science* 248 (6 April 1990), 44–48.

3. Immanuel Kant, *Groundwork of the Metaphysic of Morals,* trans. and analyzed by H. J. Paton (New York: Harper & Row, 1956). Kant's emphasis on the preservation of human worth depends on the distinction of human beings as the only creatures with a free will. An extrapolation of this to embryonic, prenatal, and neonatal life would thus leave these categories of life unprotected on the basis of either rationality or free will.

4. Pope John Paul II, "The Human Person."

5. Thomas H. Murray, "The Human Genome Project: Ethical and Social Implications," *Bulletin of the Medical Library Association,* no. 83 (January 1995): 14–21.

6. John Rawls, *A Theory of Justice* (Oxford: Oxford University Press, 1971).

7. Sophocles, *Oedipus the King,* in *Sophocles I,* trans. David Green (Chicago: University of Chicago Press, 1954), 11–76.

8. Sophocles, *Oedipus the King,* 20.

9. Sophocles, *Oedipus the King,* 69.

10. John Rawls, "Social Utility and Primary Goods," in *Utilitarianism and Beyond,* ed. Amartya Sen and Bernard Williams (New York: Cambridge University Press, 1982), 159–185.

11. David, Hollenbach, *Justice, Peace, and Human Rights* (New York: Crossroads, 1988).

12. Hans Jonas, *Philosophical Essays* (Englewood Cliffs, N.J.: Prentice-Hall, 1974).

13. Alex Mauron and Jean-Marie Thevoz, "Germ-line Engineering: A few European Voices," *Journal of Medicine and Philosophy,* no. 16 (December 1991): 649–666.

14. Jere Surber, "Obligations to Future Generations: Explorations and Problemata," *The Journal of Value Inquiry,* no. 11 (Summer 1977): 104–116.

15. Surber, "Obligations."

16. Surber, "Obligations."

17. Tristram Engelhardt, *The Foundations of Bioethics* (New York: Oxford University Press, 1996).

18. Peter Singer, *Rethinking Life and Death* (New York: St. Martin's Griffin, 1994).

19. The more moderate position in this debate is well represented by Ronald Dworkin, *Life's Dominion* (New York: Vintage Books, 1993). Dworkin contends that politicians, philosophers, and moralists operate upon the premise that each individual holds his or her own convictions about whether a fetus is a person who possesses rights and interests, and the relationship that fetal rights have to those of pregnant women. The respective positions that are espoused in this debate can lead to stances on the human fetus that are pejorative and divisive.

Instead, Dworkin contents that our convictions concerning the human fetus actually reflect one shared notion of life's intrinsic value. He refers to this intrinsic value as "sacred or inviolable," and proposes that it is the different interpretations of life's intrinsic value that constitute the basis of our disagreements. If the controversy can be related to a shared humane conviction that is understood differently, he envisions that the realization of what we share may be more fundamental than our disagreements.

20. Plato, "Theatetus," in *The Collected Dialogues of Plato,* ed. Edith Hamilton and Huntington Cairns (Princeton: Princeton University Press, 1961), 845–919.

21. John Crosby, "Person, Consciousness," *Christian Bioethics,* no. 1 (April 2000): 37–48.

22. John Crosby, "The Personhood of the Human Embryo," *Journal of Medicine and Philosophy,* no. 18 (August 1993): 399–417.

II

THE MORAL EVENT: HOW SHOULD WE LIVE?

The increasing sophistication of genetic technology tests time-honored standards and norms. The current cultural shift from moral reasoning based upon religious beliefs to a scientific foundation for ethical decision-making reflects this trend. In part, this sea change is a consequence of the pluralism of modern society, culminating in a dearth of shared values and beliefs. Different worldviews make understanding and agreement among individuals more difficult than in earlier, less complicated societies. Confusion about the meaning, purpose, and worth of human existence is a contemporary reality. Consequently, the profound questions surrounding the moral implications of purposeful inheritable genetic modification are laborious and threatening.

Part II proposes to juxtapose the predominant scientific worldview with ethics and theology to identify core virtues, duties, and principles appropriate for a comprehensive analysis of inherited genetic modification. The mechanism for this examination is the schema of the "moral event," proposed by physician and philosopher Edmund Pellegrino. The diagram of the moral event relates the character of the agent to the act, circumstances, and consequences of the moral life. Each of these four components considers a specific aspect of moral judgment and represents a different aspect of moral reasoning (see table II.1).

Pellegrino contends that few everyday ethicists rely exclusively on one mode of reasoning in their deliberations, or utilize one ethic in

Table II.1. The Moral Event

	ACTOR	*ACTION*	*CIRCUMSTANCE*	*CONSEQUENCES*
Ethic	Virtue	Deontology	Particularity	Utility
Ethicist	Pellegrino	Jonas	Levinas	Mill
Emphasis	Character	Duty	Individuality	Benefit/Harm

Source: Adapted from Edmund Pellegrino, "Toward a Virtue-Based Normative Ethics for the Health Professions," *Kennedy Institute of Ethics Journal* 5, no. 3 (September 1995): 253–277.

their response to dilemmas. Rather, moral actors implement different aspects of virtue, deontology, particularity, and utility, depending on the moral event under consideration. The application of any one theory to the intergenerational challenges of inheritable genetic modification illustrates this predicament. A technology that is both multifaceted and intergenerational strains standard ethical theories and necessitates their expansion if they are to serve as moral guides in the genetic age. The discussion of each aspect of the moral event demonstrates that while each ethical theory makes an essential contribution, each is individually inadequate to deal with the challenges of a technique that influences our descendants directly. Suffice it to say that in many instances, it is either not clear how virtues or ethical principles apply or how theoretical ethics relates to the complexities of inheritable genetic modification.

In addition, each ethical theory entails its own inherent methodological problems. Virtue theory relies on a circular logic and lacks a specific action guide; principles that are simple enough for most people to grasp have little applicability to concrete situations and often conflict with each other. The top-down approach of both principles and duties frequently alienates parties to the deliberative process. And none of these approaches, except Jonas' ethic for a technological age, includes future persons in the ethical domain. One could review additional methodological approaches and cite other inadequacies. In the end, ethical actors do not abandon existing ethical theories but intermingle them order to address complicated issues.

Part II aligns contributions from four diverse ethical theories to address the four aspects of the moral event as they apply to inherited genetic modification. The word "align" refers to the effort to splice these theories at their points of logical affinity, while maintaining integrity at their points of divergence. Pellegrino has issued a request for such an alliance:

> I am not suggesting a feeble eclecticism, a cafeteria-style ethics, that would add a spoonful of virtue here, a principle there, and a dash of consequence in another. . . . Rather, the strength of each theory must be preserved, drawn upon, and placed in dynamic equilibrium with the others in order to accommodate the intricacy, variety, and particularity of human moral acts.[1]

In response to this summons, the following chapters analyze ethicists from diverse schools of ethics, grounded in the principle of human dignity, and place them in dynamic equilibrium. We will not embark on an extensive comparison of the work of these ethicists. Rather, our purpose is to derive from their work the characteristics of human dignity pertinent to inheritable genetic modification.

Chapter 5 concentrates on the first aspect of the moral event: the character of the moral agent. According to Pellegrino, the acting person is the foundation of all moral judgments and the primary concern in the moral life. Chapters 6–8 analyze the three remaining aspects of the moral event: the act, the circumstances, and the consequences. Chapter 6 covers the moral act and draws upon the work of twentieth-century German-Jewish philosopher Hans Jonas for its philosophical underpinning. Jonas presents a deontological/teleological approach to ethics that clearly states what "ought to go on" in order to insure the "permanence of genuine human life."[2] Chapter 7 explores Emmanuel Levinas' ethics to addresses the third facet, the circumstances of the moral event. Levinas, a postmodern philosopher, grounds ethics in the primary face-to-face relationship with the Other, who represents a weakness and poverty that issue a call to limitless responsibility. The fourth aspect, the consequences of the moral event, addresses the effect of human action and the potential to foster or undermine human well-being. In chapter 8, the utilitarian ethic of John Stuart Mill provides the foundation for the consideration of outcomes. The virtues, duties, and principles identified in these three chapters constitute the ethical axis of the Inheritable Modifications Matrix, which rests upon the characteristics of human dignity.

The concept of human dignity is frequently mentioned in ethical deliberation, but is rarely described or defined. For example, the General Conference of UNESCO adopted the Universal Declaration of the Human Genome and Human Rights and a resolution for its implementation in November 1997. Human dignity was the first principle cited; the document states that "everyone has a right to respect for their dignity" and "dignity makes it imperative not to reduce individuals to their genetic characteristics and to respect their uniqueness and diversity."[3] This stance is important, but difficult to execute without practical guidelines because it leaves the definition of dignity vague.

Historically, dignity has origins in theology and philosophy. Theologically, the notion of human dignity derives from the Genesis account in the Hebrew Bible, as discussed in chapter 4. God created humans in his image and, as such, they possess inherent dignity. Characteristics associated with beings created in the image of God are the exercise of the intellect, the love of God and other beings, and stewardship of the world and all that it encompasses. Human dignity also has roots in the secular philosophy of Immanuel Kant. The notion of dignity expressed in Kant's last formulation of the Categorical Imperative speaks of the dignity that humans have as rational agents with autonomy. The Imperative states that human beings may never be used merely as a means to an end; they must always be treated as ends-in-themselves. In the *Groundwork of the Metaphysics of Morals*, having shown that humans are self-legislating ends-in-themselves, Kant ranks humans as sovereign members of a kingdom of ends and concludes that "morality and humanity, insofar as it is capable of morality, alone has dignity."[4]

There are two common objections to citing human dignity in reference to human genetics. One is that a robust construal of dignity argues against abortion, human cloning, and the generation and destruction of human embryos for research purposes. The extension of the concept of dignity to prenatal life is unacceptable to those who reject the sanctity of life position, as discussed in chapter 3. A second is the claim that the vague appeal to human dignity is seldom helpful in a biomedical context. To tackle the first concern is far beyond the scope of this book. The purpose of the following chapters is to address the latter objection by identifying the components of human dignity that are operational in genetic medicine.

NOTES

1. Edmund Pellegrino, "Toward a Virtue-Based."
2. Hans Jonas, *The Imperative of Responsibility: In Search of an Ethics for the Technological Age* (Chicago: University of Chicago Press, 1984).

3. UNESCO, "General Conference of UNESCO Adopts Universal Declaration on the Human Genome and Human Rights and a Resolution for Its Implementation," *International Digest of Health Legislation*, no. 49 (1998): 417–421.
4. Immanuel Kant, *Groundwork.*

5

The Agent: How Can We Be Good?

Wisdom and goodness to the vile seem vile.

—William Shakespeare

For a virtue ethicist, the primary question in the genetic age is: "Who will we become as we manipulate our genetic makeup?"[1] Virtue is the earliest and most common notion in ethical theory. It centers on the character of the acting person, including purpose, intentions, and disposition. The normative standard in virtue ethics is the good person. Historically, the cultural perception of "the right and the good," toward which human behavior should aim, focused on a virtuous person whom society emulated as worthy of imitation. The individual of exemplary character embodied those virtues that others in that culture should strive to possess. However, since the normative standard of virtue was commonly quite subjective, this method of character assessment often presented problems. In classical Greek ethics, Aristotle consistently used Pericles as the example of virtuous character, but this did not mean that all Greeks agreed with his choice. It was entirely possible that Athenians differed in their view from Spartans who, in turn, varied from Olympians. In fact, Aristotle acknowledged the difficulty in designating a virtuous person and recognized that the whims of popular culture in his homogeneous society were often at odds. Aristotle's struggle is magnified in our contemporary pluralistic culture.

VIRTUE ETHICS

The origins of the concept of virtue appear in the writings of Plato and Aristotle and later in the texts of the Stoics and Epicureans. Plato introduced the notion of virtue in the *Meno,* when Meno asked Socrates:

> Can you tell me, Socrates—is virtue something that can be taught? Or does it come by practice? Or is it neither teaching nor practice that gives it to a man but natural aptitude or something else?"[2]

These questions initiated a sustained inquiry into the meaning, structure, and acquisition of virtue that some scholars feel Socrates never resolved conclusively. There is some indication in the *Meno* that the essence of virtue consisted of "desiring the good," but the concept of the good remained an abstract idea.

Aristotle dealt with the concept of virtue differently than Plato. He approached the life of virtue from a more practical orientation than his teacher did. He retained the idea of desiring the good promulgated by Plato, but defined it as more than a transcendent idea. For Aristotle, the "human good turns out to be activity of the soul in conformity with excellence, and if there is more than one excellence, in conformity with the best and most complete."[3] This notion of the good entailed a strong concept of *telos,* or the end toward which things tend. Because of this end-oriented focus, Aristotle's ethics are called teleological, meaning that he shaped ethics around the idea of an ultimate purpose or end, defined as the achievement of the human good.

From this orientation, Aristotle taught that virtue is excellence of human character acquired by practice: "The virtues on the other hand we acquire by first having actually practiced them, just as we do the arts." He is clear that the virtues are a matter of conduct and that they result from a disposition to virtue acquired from childhood. In addition, continual coaching in the acquisition of virtue is "of supreme importance." As such, virtues are not divinely conferred, but are acquired through habitual practice.

> In order for an act to be virtuous, the agent must be in a certain condition: First he must act with knowledge; secondly, he must deliberately choose the act, and choose it for its own sake; and thirdly the act must spring from a fixed and permanent disposition of character.[4]

As such, the classical concept of virtue was a state of character or a disposition, chosen deliberately and rationally for its own sake. It denoted a predictable disposition for a person to "perform its function well" in all circumstances.

Christian Virtue Ethics

Augustine christianized virtue ethics in the fourth century and Thomas Aquinas further expanded the notion in the thirteenth century. The subsequent amalgamation of Greek and Christian doctrines of virtue constituted the dominant shape of ethics for over two millennia. Augustine and Aquinas presented the notion of virtue as "traits of character, as perfections of human conduct of a moral and intellectual kind."[5] In addition, both regarded the cardinal virtues (courage, prudence, temperance, justice) handed down by the Greeks as insufficient for the pursuit of the Christian life and integrated the theological virtues of faith, hope, and charity into their moral philosophies. For Augustine, this realization arose from his belief that the Platonic idea of the knowledge of "the good" was insufficient to prompt the choice of the good life. Only the realization of a supreme God who sustains the world and its inhabitants through grace provided sufficient motivation to pursue the good. To motivate the choice of good over evil, Augustine believed that the will must be involved. He contended that happiness resulted from the pursuit of life with God and the practice of the theological virtues.

Aquinas, theologian and master of sacred doctrine, defined virtue as that which makes one good and renders one's activity good. Like Aristotle, Aquinas stated that human virtue is the habit of perfecting a human being so that

> it acts well. There are two principles of human action, namely intellect or reason and appetite. . . . Hence any human virtue must be perfective of one or the other of these principles. If it is perfective of speculative or practical intellect so that a person acts well, it is an intellectual virtue; if it is perfective of the appetitive part, it is a moral virtue.[6]

He identified the end of the moral virtues as the "good for humans." In order to know whether something or someone was good, one must know one's purpose. It was at this juncture that the intent of virtue for Aquinas deviated from the Greeks and joined Augustine. Both Aquinas and Augustine believed that the "good for humans" was a supernatural end: humans were created to be in union with God. The intent of the Christian virtues was to empower persons to attain the ultimate purpose of union with God: "faith orients to that end, hope enables the Christian to stay the course, and charity attaches the Christian to God."[7]

The acquisition of the theological virtues differed from the pursuit of the cardinal virtues in that the former were gifts from God, not products of habitual practice. The Christian virtues disposed humans to a purified form of earthly life, the source of which was not human effort but the infusion of grace. As such, sin was not simply vice or the antithesis of virtue. It was an offense against God that separated humans from the divine life by destroying the channel of grace and the possibility of a life of virtue. Only repentance negated the effect of sin and restored the Christian to a state of grace.

Contemporary Virtue Ethics

Plato, Aristotle, and Aquinas instructed a homogeneous citizenry who espoused a common, albeit exclusive, view of the good. Contemporary virtue ethicists do not have this luxury. Modern pluralism by definition negates a shared notion of the good for humankind. Virtues and vices are presently interchangeable, with some qualities considered a virtue by some and a vice by others. For example, some groups categorize pride as a vice analogous to arrogance, whereas feminist ethicists rank pride as a virtue analogous to self-respect. This disparity in beliefs caused many philosophers to abandon or deny the quest for the possibility of a contemporary virtue ethic.

However, there is a concurrent resurgence of interest in virtue ethics. Twentieth-century philosopher Alasdair MacIntyre recaptured the Aristotelian notion of virtue in contemporary terms. For MacIntyre,

> a virtue is an acquired human quality the possession and exercise of which tends to enable us to achieve those goods which are internal to practices and the lack of which effectively prevents us from achieving any such goods.[8]

This definition uses the concept of practice in an attempt to avert the difficulty of defining the good. A practice is a "complex socially established cooperative human activity through which goods internal to that form of activity are realized."[9] The range of practices includes those activities that sustain communities and provide historical and familial contexts for individuals; its definition encompasses fields as diverse as mathematics, portrait painting, and farming. For example, painting a room is not a practice, but interior design is; playing catch with a child is not a practice, but the game of baseball qualifies. The criteria that must be met are established standards of excellence and accepted rules that have the achievement of goods as an end point.

The concept of practice provides a means to define virtue and implement a common objective among moral agents. The present erosion of established traditions and moral consensus requires the designation of a specified practice with a shared sense of the good in order to determine a purpose and identify pertinent virtues. Otherwise, the possibility of reestablishing a normative role for virtue ethics in the absence of a shared philosophical or theological foundation is remote. For the purposes of this analysis, the designated practice is human medical genetics, with its professionals participating in the development and potential application of inheritable genetic modification.

THE MORAL COMMUNITY

Pellegrino defines a moral community as "one whose members are bound to each other by a set of commonly held ethical commitments and whose purpose is something other than mere self-interest."[10] The moral purpose resides in the "nature of the activity" undertaken or in the "commitments" freely espoused by the participants in the endeavor. As such, disregard for agreed-upon "commitments" entails disassociation from the moral community. A moral community differs from others that entail amoral or immoral pursuits on the basis of espoused goals. According to Pellegrino, it must embrace goals that are morally defensible. By definition, a moral community uses its power for good and exemplifies meritorious values. The designation of a moral community provides the foundation to identify virtues germane to the appraisal of inheritable genetic modification. The moral community in the field of human medical genetics includes professionals working at all levels: administrators, research scientists, physicians, and genetic counselors.

The Moral Facets

To substantiate the designation of human geneticists as a moral community, it is necessary to identify the facets that allow its participants this status. Pellegrino cites four aspects that designate medicine as a moral community:

> the obligatory nature of the knowledge considered, the characteristic of significant moral decisions, the complicity of the professional, and the inequality in the professional relationship that constitute ethical commitments which surpass mere self-interest.[11]

An analysis of the practice of human genetics clearly reveals the presence of these four parameters. First, the nature of human genetic knowledge entails intrinsic characteristics that generate moral obligations for those involved. It is knowledge about the primordial material of human inheritance and development, and, as such, calls for accountability for its development and use. Second, the qualities of decisions concerning inheritable genetic modifications are significant in that they are intergenerational and have societal effects. Changes in the human germ line affect those as yet unborn, in addition to impacting the social status of others who carry genes designated harmful or defective. Third, the moral responsibility for the results of genetic research is a component of professional endeavor. As gatekeepers for the creation and utilization of genetic knowledge, the geneticist is obligated to employ her knowledge for the good of humankind.

The fourth moral aspect of human genetics involves the nonscientist, specifically the patient. Given the complexity of genetic information and the profundity of the choices involved, the patient is largely at the mercy of the geneticist. The discrepancy of knowledge that exists between professional and patient is tremendous and inescapable, and requires a reliance on the trustworthiness of the professional. The power wielded by the genetics professional culminates in obligations that are morally significant. The presence of these four facets—the nature of genetic knowledge, the characteristics of genetic decisions, the moral complicity of the professional, and the inequality in the patient-professional relationship—delineates human genetics as a moral practice.

THE GOOD OF HUMAN GENETICS

Pellegrino identifies three necessary ingredients for the development of any normative ethical framework for the practice of medicine based in virtue:

> (1) a theory of medicine to define the *telos,* the good of medicine as an activity; (2) a definition of virtue in terms of that theory; and (3) a set of virtues entailed by the theory to characterize the "good" health professional.[12]

Our first task is to identify the good of human medical genetics and determine the purpose of the practice. The second requisite, a definition of virtue, was discussed earlier. The final criterion, a set of virtues specific to the good geneticist, will also need our attention.

The basis upon which to tackle the first component, the good of human genetics, is the "goods internal to the practice." David Hull's *Science as a Process* cites the three goods of the practice of science as

> a desire to understand the world in which we live, the allocation of responsibility for one's contributions (both credit and blame), and the mutual checking of these contributions; in short, *curiosity, credit, and checking.*[13]

These goods also apply to human genetics, since it constitutes a subset of the general practice of science.

The first good, curiosity, is innate to human beings. The curiosity to understand DNA and its function in the development and maintenance of life is natural. It constitutes the pursuit of knowledge for its own sake. Historically, healthy societies are those in which the arts and sciences thrive. Individuals pursue areas of artistic and technical interest as a personal and common good. The curiosity aroused by the observable patterns of genetic inheritance stimulates the human appetite for understanding. The freedom to pursue curiosity is a necessary component of human genetic research. The second good of human genetics, credit, emphasizes the importance of recognition in genetic research. The most important acknowledgment that scientists can receive for quality work is its use by other scientists working in the same area. In addition to recognition, the element of credit encompasses the responsibility for harm or good ensuing from one's endeavors. The third good comprising a theory of human genetics is checking, which entails the verification of reported results by peers in subsequent research or clinical applications. Fabrication or falsification of scientific findings is unethical and can have a tremendous ripple effect on countless related projects. Checking is an example of a practice of science where virtue and professional self-interest merge. It is in the interest of the professional to avoid dishonesty and the likelihood of exposure.

The Purpose

The purpose of human genetics originates from the analysis of the "goods internal to the practice" discussed above. The practice of human genetics embraces two realms: the immediate and the collective. The immediate realm consists of the daily individual endeavors of the geneticist; the collective realm involves the ramifications of individual actions on society at large. The immediate purpose of the geneticist is the responsible advancement of scientific knowledge and the development of clinical applications. A "good" geneticist wants to practice interesting and safe science on a daily basis. The collective purpose of the geneticist is to encourage the application of genetic knowledge for the good of humankind. This two-faceted purpose evolves directly from the three goods of curiosity, credit, and checking. Scientists are curious about the world in which they live, and they care about the knowledge that they generate.

The good that summarizes the dual purposes of human genetics is: *The good of human genetics is the responsible discovery and application of genetic knowledge to prevent, cure, or care for those with genetic disorders and promote general well-being.* This statement is an encapsulation of the goods internal to the practice of molecular genetics, coupled with its designation as a moral community. The tenet of responsible discovery flows directly from the curiosity to understand human genetics, plus the obligation to substantiate the knowledge produced. The amelioration of genetic disorders and the promotion of health are means of obtaining credit for one's work.

This approach is primarily teleological without being consequentialist, that is, it does not employ reasoning that limits the good to the calculation of quantitatively measurable consequences. It is teleological in that it shapes the ethics of human medical genetics around the acknowledgment of a stated purpose. The description of the purpose of human genetics provides foci for ethical discernment. A choice of action is right or good based on the degree to which it manifests character traits that promote the stated purpose of human genetics. Conversely, geneticists that pervert the purpose of human genetics are unethical. The virtues described below provide the foundation for the character of a moral genetics professional.

VIRTUES OF THE "GOOD" GENETICIST

The final component of a normative approach to virtue theory is a summary of the virtues that characterize the moral geneticist. The goods internal to the practice of human genetics and the purpose of the profession dictate the virtues particular to this area. This list of virtues is not all-encompassing. When considering the practice of human genetics, the scope of which has been likened to the Apollo moon launch, the Manhattan Project, and the exploration of the American western frontier, the register of virtues could be endless. The four virtues that are critical to the goods and purposes of human genetics are wisdom, humility, justice, and integrity. They are not listed in order of preference; an amalgamation of them is necessary for the geneticist to carry out her tasks admirably. For Aristotle, the end of virtue is human excellence, which then "will be the disposition that makes one a good man and causes him to perform his function well."[14]

Wisdom

Wisdom is an intellectual virtue that is linked to truth, knowledge, and reason. For Socrates, virtue was knowledge and to be wise was to be good. Aristotle defines wisdom as the disposition to deliberate correctly on what is good or bad for humankind. Aristotle makes a clear distinction between philosophical wisdom and practical wisdom. Both are intellectual virtues, but they differ in function. For Aristotle, philosophical wisdom is the consummate form of knowledge: it is "scientific knowledge, combined with intuitive reason, of the things that are highest by nature."[15] At first glance, it seems obvious that philosophical wisdom is the key virtue for a geneticist to practice, since the knowledge dealt with is among the most profound in science. However, further examination reveals that practical wisdom is also especially pertinent to the challenges posed by the genetic age. Practical wisdom is the wisdom of action; it directs how to act well in order to live well. Aristotle describes Anaxagoras, Thales, and others like them as men who

> have philosophic but not practical wisdom, when we see them ignorant of what is to their own advantage, and why we say that they know things that are remarkable, admirable, difficult, and divine, but useless; because these sages do not know the things that are good for human beings.[16]

Given the power and complexity of genetic knowledge, the gatekeepers of this arena need philosophical wisdom in the discernment of the good for human beings and practical wisdom to show the way.

Practical wisdom is the ordering virtue for Aristotle; it guides the exercise of all other virtues. Frequently in the pursuit of the moral life, a practical appraisal of competing virtues, principles, and duties is necessary; practical wisdom is the virtue that administers these decisions. It is both an intellectual and a moral virtue since it addresses "things human and things about which it is possible to deliberate." Aristotle defines practical wisdom as the habit "to deliberate well about what is good and advantageous for himself, not in some one department . . . but what is advantageous as a means to the good life in general." To deliberate well refers to "correctness of thinking."[17]

The ability to deliberate well facilitates the procurement of the goods of human genetics: the pursuit of knowledge and the potential to impact for good. Those moral agents that practice "correct thinking" discern the applications of genetic knowledge that function to correct genetic disorders and promote health. Finally, and most important, the geneticist exercising practical wisdom has the preeminent opportunity to draw the line to distinguish between those procedures that thwart or reinforce the purpose of human genetics. Any research or clinical applications that transgress the responsible discovery and application of human genetic knowledge to ameliorate genetic disorder and promote health also transgress the moral boundaries of the profession.

As we are propelled into the postgenome era, we encounter new territory, posing questions that have never been posed before. The virtue of wisdom is essential if the geneticist is to make difficult distinctions concerning right and wrong, good and evil. Jonas succinctly expresses the need for the formation of character steeped in the practice of wisdom:

> [W]e are constantly confronted with issues whose positive choice requires supreme wisdom—an impossible situation for man in general, because he does not possess that wisdom, and in particular for contemporary man, who denies the very existence of its object: viz., objective value and truth. We need wisdom most when we believe in it least.[18]

Humility

Encompassed in the virtue of practical wisdom is a virtue that is overlooked and particularly unpopular in the modern age. This is humility, the virtue that recognizes what is beyond the limits of human wisdom. Jonas calls for a revitalization of the virtue of humility in light of the tremendous power that human genetic technology has the potential to manifest. The virtue that Jonas describes is a "new kind of humility" that correlates with the magnitude of human control over our genome. New humility differs from former interpretations of humility in that it focuses not on the insignificance of human ability but on the extent of our power. Humility in this new role emphasizes the "excess of our power to act over our power to foresee and our power to evaluate and to judge."[19] The nearly unlimited potential to modify our environment and ourselves, coupled with a limited ability to predict the ramifications of our actions, requires the rediscovery of humility.

Justice

Justice is both a principle and a virtue; it is the only virtue that is a good in itself, because it comprises a good done to others. It was discussed as the Rawlsian principle of justice in chapter 4. The virtue of justice is equivalent to the principle of justice in its root meaning: justice is the habit of giving another member of society her due. The pursuit of this virtue spans the centuries from the ancient Greeks to the present. Plato scrutinized the virtue of justice in *The Republic* from the standpoint of paying debts, actions toward friends and enemies, and goodness of attitude. He never definitively resolved his quest for the meaning of justice, but concluded that it is a fitting virtue since it is directly connected to human well-being. Aristotle stated that justice is a complete virtue, for in justice is summed up the "whole of virtue"; this is because it is the boundary that defines the other virtues and allows them to function together. Aristotle defined the just person as one who "when making distributions between himself and another, or between two others, will not give himself the larger and his neighbor the smaller share of what is desirable (and vice versa in distributing what is harmful)."[20]

Aquinas expanded the breadth of the virtue of justice through the Christian notions of love and charity. Justice augmented by love transcends legal justice and brings an expanded significance to giving the other her due. The Christian virtue of justice embraces the notion of special concern for the marginalized, the poor, the sick, and the needy. It calls the virtuous person to express the communal aspects of justice in concrete acts of benevolence toward specific individuals. As such, it is neither an abstract concept nor a strict calculation of gains and losses, but a way of love generated by faith and expressed in charitable action.

Echoing the Christian definition of charitable justice from a particularistic perspective, Levinas contends, "Does not justice consist in putting the obligation with regard to the Other before obligations to oneself, in putting the Other before the Same?"[21] For Levinas, the possibility of a theory of justice or a just society stems from the fundamental structure of a relationship with the Other. The limitless responsibility that Levinas finds at the core of the relationship with the Other culminates in justice. Although Levinas grounds justice in the personal relationship, he provides a way to understand it on the collective level. He moves from relational justice to collective justice through his account of the third party.[22] For Levinas, the presence of the third party is not an empirically identifiable event, but an occurrence that implies the possibility of innumerable others. It is a possibility that precludes an insular relationship between the self and the Other. The third party calls attention to the presence of the Other as not merely for the self but for the neighbor as well. The presence of the third party opens up a broader perspective that provides a foundation for society and an obligation for social justice. The third party makes possible an acknowledgment of social awareness and allows Levinas's nonuniversalistic ethics to apply to the concrete concerns of social justice. Justice begins with the Other and culminates with the awareness of the privilege that the Other has relative to the self. Justice precedes personal freedom, and the Other is the measure of one's injustice. Extending Levinasian justice to human genetics, the analysis of any procedure would first consider the obligation to the Other: "[I]t puts into question the naive right of my powers."[23]

Many proffer accusations against the Human Genome Project of transgressions of justice, such as the project's cost and scale. It is certainly possible that the HGP saps resources from smaller projects that aid far greater numbers of needy individuals. Historically, new technologies magnify rather than diminish economic inequalities (at least initially), although new vaccines based on functional genomics already benefit many. That said, most of the prominent research endeavors of the HGP and biotechnology companies relate to the diseases of wealthy nations. There is scant interest in the truly intractable scourges of the third world that generate little economic reward. One can easily envision access to curative genetic services for the rich, while the poor continue to suffer the ravages of genetic disease due to a lack of the ability to pay. Yet it is precisely to the plight of the poor to which the virtue of justice pertains. The just human geneticist exercises a "preferential option" toward those most needy in society.

> It is, in fact, precisely to the "losers" in the natural lottery—the sick, the poor, the outcast—that Christ addressed his personal ministry, his healing miracles, and his Sermon on the Mount. This is the basis for the preferential option for the poor that inspires the best Christian institutions.[24]

The just human geneticist acts to ensure that all humans, especially the most vulnerable, benefit from advances in genetic medicine.

Integrity

The virtue of integrity is the habit of acting in a principled and honorable fashion in relations with others; it is fidelity to trust. Trustworthiness is an achievement, not a mandate or mere compliance to the letter of the law. A geneticist with integrity does good science, honestly reports and interprets data, and pursues ethical rigor in her work. The future of any research enterprise is contingent on the central role of trust among the public, the funding source, and the geneticist. The practice of integrity in science inspires the public trust necessary for the realization of scientific advancement.

The antithesis of the virtue of integrity is the vice of self-interest. Self-interest is a normal human incentive and is not inherently immoral. Only when its pursuit abrogates the goods of genetics research or interferes with doing one's work in a trustworthy fashion does it become a vice. The inclusion of the vice of self-interest in the discussion of the virtue of integrity is not meant to cast them as polar opposites, like good and evil. Twentieth-century Christian ethicist Reinhold Niebuhr advocates the ability to acknowledge the taint of self-interest in every moral position, including one's own.[25] He asserts that human action almost always includes a degree of self-interest. Niebuhr maintains that rather than attempting to obliterate it, the more effective approach is to recognize and control it.

The inducement to the vice of excessive self-interest arises partly from the conversion of human genetics from a scholarly pursuit within a university setting to a profitable business enterprise. This metamorphosis exposes the goods internal to the practice of genetics to the values of business that include competition, ownership, and investment.

In the absence of the habitual practice of integrity, the values of business could thwart the purpose of genetics research and encourage the magnification of self-interest to unethical levels. The academic community is not free from similar enticements. The potential for scholarly awards, power, and prestige is equally alluring to some as financial rewards are. As Aristotle insisted, virtues are acquired by practice. The final safeguard of the morality of research and the responsible application of human genetic knowledge is the integrity of human geneticists in both the private and public sectors. The development of the virtue of integrity requires rigor and dedication.

THE DIFFICULTY AND MERIT OF VIRTUE ETHICS

One of the designated tasks of this chapter is to demonstrate that virtue ethics, while vital, is insufficient in itself to direct the utilization of human genetic technology. Pellegrino identifies three difficulties with the practice of virtue as it applies to moral action. The first lies in its circular logic, which holds that "the good is that which the virtuous person does and the virtuous person is the person who does what is good for humans."[26] The current lack of consensus as to what defines the good compounds this problem. The second difficulty is that of supererogation, or going beyond what is required. In a society based on legalism and individualism, the only duty is to avoid encroachment on another's liberty. It is beyond one's duty to require any more than a minimalist approach to moral conduct based in virtue. The third difficulty lies in the lack of a definitive method in virtue ethics to establish action guidelines, because virtue relies on the character of the virtuous person rather than on a set of directives.

The lack of norms presents a particular challenge in virtue ethics. Aristotle taught:

> Virtue is a state of character concerned with choice, lying in a mean, i.e. the mean relative to us, this being determined by a rational principle, and by that principle by which the man of practical wisdom would determine it. Now it is a mean between two vices, that which depends on excess and that which depends on defect.[27]

A person is virtuous when he or she chooses the mean between the two extremes that demarcate the contrasting vices:

> By the mean of a thing I denote a point equally distant from either extreme, which is one and the same for everybody; by the mean relative to us, that amount which is neither too much nor too little, and this is not one and the same for everybody.[28]

For example, a courageous soldier chooses the mean between the vices of cowardice and recklessness before deciding upon a path of action. Aristotle states that the habitual choice of a mean is determined by a standard that is intuited rationally. This introduces subjectivity into the determination of the mean and ambiguity into the definition of virtuous conduct. As defined, the mean depends upon the interpretation of each moral actor. Aristotle is also quite clear that there are actions for which there is no mean. This is due to the fact that some actions are always wrong, for example, "adultery, theft and murder."

Having considered the above limitations, one might wonder why the ethic of virtue is indispensable to the deliberation concerning inheritable genetic modification or any other normative question. The merit in virtue ethics is its practical application by the moral actor. The moral actor is never separate from action. The pursuit of the good for humankind ensues from the character of the agent, more often than from the application of rules or the consideration of consequences. In the final analysis, even the most stringent laws cannot preclude the pursuit of immoral or harmful human genetic research somewhere in the world.

Virtuous character is the basis of the moral life. This analysis demonstrates that the virtues germane to the purpose of human genetics can be specified within a moral community, despite differing individual worldviews. As theologian Peter Bryne explains:

> Courage, justice, friendship, the power of thought and the exercise of intelligence, [and] self-control are the disciplines that in the abstract ideal are the essential Aristotelian virtues, although the concrete forms that they take vary greatly in the different socially conditioned moralities. The virtues of splendid aristocratic warriors are not the same as the virtues of a Christian monk; but are not merely different. Each of the two ways of life demands courage, fairness or justice, loyalty, love and friendship, intelligence and skill, and self-control.[29]

Bryne contends that while disparate theologies and worldviews culminate in "different concrete realization of the human good," there is a "minimal, common starting point" from which debate is reasonable.

The combined virtues of wisdom, justice, humility, and integrity comprise a starting point and a reasonable basis upon which to model the character of the genetics professional. The identification of virtues specific, but not restricted, to the character of the geneticist illustrates the possibility and merit of virtue in the pursuit of the professional life. The following chapters splice virtue ethics with duty, particularity, and utility in an effort to propose additional components necessary for a matrix addressing the ethical applications of inheritable genetic modification.

NOTES

1. James F. Keenan, "What Is Morally New in Genetic Manipulation?" *Human Gene Therapy,* no. 1 (1990): 289–298.
2. Plato, "Meno," in *The Collected Dialogues of Plato,* ed. Edith Hamilton and Huntington Cairns (Princeton: Princeton University Press, 1961), 353–384.
3. Aristotle, *Nichomachean Ethics,* trans. H. Rackham (Cambridge, Mass.: Harvard University Press, 1994).
4. Aristotle, *Nichomachean.*
5. Augustine, *The City of God,* trans. Marcus Dods (New York: The Modern Library, 1993).
6. Thomas Aquinas, *Summa Theologica,* trans. the Fathers of the English Dominican Province (London: Burns Oates & Washbourne Ltd., 1953).
7. Edmund Pellegrino, "Toward a Virtue-Based Normative Ethics for the Health Professions," *Kennedy Institute of Ethics Journal* 5, no. 3 (September 1995): 253–277.
8. Alasdair MacIntyre, *After Virtue* (Notre Dame, Ind.: University of Notre Dame Press, 1984).
9. MacIntyre, *After Virtue.*
10. Edmund Pellegrino, "The Medical Profession as a Moral Community," *Bulletin of the New York Academy of Medicine,* no. 66 (May–June 1990): 221–232.
11. Pellegrino, "The Medical Profession."
12. Pellegrino, "Toward a Virtue-Based."
13. D. Hull, *Science as a Process* (Chicago: University of Chicago Press, 1995).
14. Aristotle, *Nichomachean.*
15. Aristotle, *Nichomachean.*
16. Aristotle, *Nichomachean.*
17. Aristotle, *Nichomachean.*
18. Hans Jonas, *Philosophical Essays* (Englewood Cliffs, N.J.: Prentice-Hall, 1974).

19. Jonas, *Philosophical.*

20. Aristotle, *Nichomachean.*

21. Emmanuel Levinas, *Collected Philosophical Papers,* trans. Alphonso Lingis (Boston: Martinus Nijhoff Publishers, 1987).

22. Emmanuel Levinas, *Otherwise Than Being or Beyond Essence* (Boston: Martinus Nijhoff Publishers, 1981).

23. Levinas, *Collected Philosophical.*

24. Edmund Pellegrino, *The Christian Virtues in Medical Practice* (Washington, D.C.: Georgetown University Press, 1996).

25. Reinhold Niebuhr, *The Nature and Destiny of Man,* vol. 2 (New York: Charles Scribner's Sons, 1942).

26. Pellegrino, "Toward a Virtue-Based."

27. Aristotle, *Nichomachean.*

28. Aristotle, *Nichomachean.*

29. Peter Bryne, *The Philosophical and Theological Foundations of Ethics* (New York: St. Martin's Press, 1992).

6

The Act: What Is Our Duty?

> A sense of duty pursues us ever. It is omnipresent, like the Deity. If we take to ourselves the wings of the morning, and dwell in the uttermost parts of the sea, duty performed or duty violated is still with us, for our happiness or misery. If we say the darkness shall cover us, in the darkness as in the light our obligations are yet with us.
>
> —Daniel Webster

As we stand at the threshold of the genetic revolution, Jonas exhorts that "it is fitting that man should pause a moment for fundamental reflection."[1] Jonas proposes an ethic for the technological age that focuses on the obligation of ethical agents to ensure the future existence of humankind. While Jonas endorses virtue ethics as necessary "old prescriptions of the 'neighbor' ethics—of justice, charity, honesty, and so on," he believes that virtue is limited to daily human interaction and contends that genetics' ability to impact the arena of collective action overshadows the sphere of immediate interaction.[2]

In the purview of genetics, the ethical actor, the act, and the effects are no longer in proximate contact. The actions of the geneticist impact society and future persons specific in ways that were previously impossible. Jonas maintains that the enormity of the power brought to bear by the changed potential of human action necessitates a new dimension for ethics never before required. In response, he proposes an ethic that is duty-based but goal-oriented. He moves the focus of ethics from the proximate range of contemporary action to the remote realm of future responsibility. This is a departure from previous ethics that focused on the present, since until now, all moral agents shared in a common here and now.

Chapter 6 analyzes the moral act through the lens of Jonas' ethic, which was specifically created to include future persons in the ethical domain. The analysis of the moral act begins with Jonas' value-oriented approach to ethical duty, which he grounds in the innate worth of human existence. From this foundation, Jonas' Imperative of Responsibility presents the formal concept of the ethical agent's duty to ensure the continuation of human life. The obligation of *responsibility* focuses on the assessment of actions in the matrix. This chapter concludes with a comparison of Jonas' Imperative to its Kantian ancestor, the Categorical Imperative. The juxtaposition of the two imperatives provides a way to consider an ethical boundary for inherited genetic modification.

SHOULD THERE BE HUMANS?

Advances in biotechnology occurred simultaneously with a societal devaluation of the sacred and an aversion to divine revelation. The expulsion of religious language from the public debate, coupled with the flourishing of biotechnology, resulted in an obliteration of the traditional beliefs about humankind and an ensuing nihilism in existing norms. Jonas observed that in the subsequent vacuum, the question of human beings and whether we "ought to be" passed to the domain of philosophy. Jonas accepted the challenge and focused the ontological argument upon ethics. To address this task, he chose to query of existence, "Is it worth being?" rather than to ask Leibniz's question of being versus nothingness. The choice of the former question allows Jonas to focus on a justifying norm derived from the value perceived in human existence. Value, once demonstrated, is an element with the potential to ground a claim for being, because the presence of value relegates the continuation of being to the domain of duty. In addition, posing the "worth in being" question avoids the issue of authorship in creation, thus detaching it from matters of faith. Once Jonas establishes value in being, he can demonstrate the superiority of being over nothingness.

The ensuing task is to explore his construal of the objectivity of value. Only from an objective standard of value can a statement that someone or something "ought to be" emanate. "Ought" is the grammatical signal for ethics and for ethical obligation. Jonas derives the objectivity of value from the presence of ends or purpose. As stated previously, an end is that purpose for which matter exists. The presence of ends entails the suitability of matter for those aims and the ability or lack of ability to attain those purposes. From this premise, Jonas proposes the capability to achieve a purpose as a means for determining the presence or absence of good in matter. The example of purpose that Jonas uses is the hammer, which is a tool that has as its end the function of hammering. He maintains that one can judge, relative to the hammer's ability to drive a nail or a spike, whether it is a good or bad tool.

The final question that Jonas must answer is: Once one demonstrates that purpose is present in nature, does this legitimize purpose and thus ground obligation? To establish obligation merely on the basis of the presence of purpose in nature would be to cross the forbidden gulf between "is" and "ought." Jonas responds with a distinction between a particular purpose and the presence of "purposiveness" as the ontological property of an entity: "We can regard the mere *capacity to have* any purposes at all as a good-in-itself, of which we grasp with intuitive certainty that it is infinitely superior to any purposelessness of being."[3] Jonas positions the superiority of purposiveness over purposelessness as a *first principle,* or one that does not require additional substantiation.

Thus, Jonas posits the reality and efficacy of purposiveness in nature as a fundamental affirmation of being over nothingness. "In every purpose, being declares itself for itself and against nothingness."[4] The important distinction is not so much between something and nothing, but rather in the disparity between purpose and indifference. The first "yes" to being comes from itself, in its interest and concern with existence. Subsequent value derives from the maximization of purposiveness. The more intense the purposiveness, the more emphatic is the affirmation of being. The presence of purposiveness in being bestows the corresponding designation of "truly worth the effort." The presence of value confirms worth in being, and as such issues a moral ought for the preservation of humankind. This moral ought grounds Jonas' imperative.

THE OBLIGATION OF RESPONSIBILITY

The responsibility delineated in Jonas' imperative is substantive. The traditional notion of responsibility calls one into account for past and present performance. In contrast, substantive responsibility focuses on the future determination of what will be done. "By its command I feel responsible, not in the first place for my conduct and its consequences but for the *matter* that has a claim on my acting."[5] Substantive responsibility is not merely the formal idea of duty, but also includes the sentiment of responsibility. The duty to act responsibly springs from the emotion emanating from a worthy goal: the possibility of good in the world. For Jonas, ethics has two integral facets, a subjective side dealing with emotion and an objective component dealing with reason. The two facets are mutually complementary and, as such, constitute the ethical domain. Jonas contends that without the emotional component, the most convincing appeal of right is powerless to motivate a response. The reality of the power of the present action to impact the future engenders a responsibility that encompasses the sentimental.

Without the demonstration of right, a sentimental response is arbitrary and lacks justification. This explanation still admits the prospect of a purely good will whose disposition is naturally in accord with the moral law and responds without prompting. The possibility of a good will elicits a feeling of responsibility on the part of an individual whose every action influences the existence or nonexistence of being. As such, Jonas has elevated the idea of humankind's permanence in the world from the status of a general axiom to that of a moral obligation. To define this obligation, Jonas postulated the Imperative of Responsibility to consider the new type of human action that this future-oriented responsibility entails. It is a goal-oriented concept of responsibility that is dedicated to the future existence of being:

> Act so that the effects of your action are compatible with the permanence of genuine human life; or expressed negatively: Act so that the effects of your action are not destructive of the future possibility of such life.[6]

By this imperative, Jonas designates human subjects as the primary focus of responsibility. The first duty is to prohibit that which thwarts the continuation of humankind. This imperative speaks to the permanence of the "idea" of humankind, not of specific human individuals. It mandates the continuation of humankind and makes this a duty for those possessing the power to ensure or endanger that existence.

Power is present in nature, but it is restrained; in human hands, power joins with knowledge and free will and thrusts aside many of the shackles of restraint. Thus, power is the force that binds will and obligation, and moves responsibility into the forefront of ethics for the genetic age. Duty springs from power, and responsibility is a correlate of that power. Since humankind alone possesses the power to exercise responsibility, Jonas proposes the Imperative of Responsibility as the starting point of ethics. A comprehensive reading of his theory clearly extends the duty beyond humans to include all of nature. However, since the focus of this work is human genetics, our discussion of Jonas is restricted to humans.

Responsibility and Ethics

Until the advent of the genetic age, the power of humankind was confined to the proximate present. Actions resulted in changes that were limited in scope and duration. The knowledge required to guide such action was proportionate to the scope of human power. It necessitated knowledge that normal individuals had available. Kant was unequivocal in his assertion that "human reason can, in matters of morality, be easily brought to a high degree of accuracy and completeness even in the most ordinary intelligence."[7] In other words, one need not be a philosopher or a scientist in order to make good moral judgments. Jonas argues that biotechnology calls that supposition into question. The power that humans orchestrate through genetic alteration is immense and requires commensurate knowledge.

Presently, humans can fabricate themselves, their community, and some aspects of future generations. In the genetic age, the role of "man as maker" takes on a new dimension. "If the realm of making has invaded the space of essential action, then morality must invade the realm of making, from which it had formerly stayed aloof, and must do so in the form of public policy."[8] Once the level of human action escalates to the collective and future whole, a duty to evaluate the outcome of human action becomes essential. The knowledge of the ramifications must correlate with the causal scale of action. Jonas contends that our predictive knowledge lags behind the technical knowledge that enables our ability to act, and that this reality assumes ethical importance.

To balance the technical power requires a proportionate knowledge that Jonas calls predictive knowledge. He contends that human prognostication is inadequate when compared to the technical knowledge that empowers the genetic age. The discrepancy between human capability to wield unprecedented power through inheritable genetic modification and the inadequacy to predict the ramifications of those actions creates a moral quandary for which traditional ethics is unprepared.

VALUE AND DEONTOLOGY: THE IMPERATIVE COMPARISON

Jonas proposes the Imperative of Responsibility as an ethic for the technological age. It is reminiscent of the Kantian Categorical Imperative: "Act so that you can will that the maxim of your action should become universal law."[9] The Kantian Imperative has its source in reason and requires the consistency of free will with itself and that of others. For Kant, an action is moral if it can be willed universally without self-contradiction. Moral reflection is rational and logical, and a judgment of moral necessity constitutes an obligation. The similarities between Jonas and Kant are obvious: both are duty-based and issue imperatives in the form of a mandate or command. Their differences are also evident. Kant denies that sentiment supplies motivation for response to the moral law, and for him temporal objects do not evoke sentiment. Only the idea of duty inspires a sentiment of reverence for the law. Reason itself becomes the source of the affections, and what ought to be issues from reason.

In addition, Kant differs from Jonas in that the Categorical Imperative is addressed to individual decision-making: its universalizability is only a theoretical possibility. Jonas addresses the Imperative of Responsibility to societal action, or public policy, rather than to personal judgments. His test of morality is the future collective ramifications of human action rather than the morality of the individual act. He predicates morality upon whether or not the results of human action ensure the permanence of humankind on this planet. As such, the consequences considered are not hypothetical but substantive. They do not affect a theoretical community but the collective whole of humankind.

Value

Some philosophers would contend that the most basic difference between these ethicists lies in the presence of Jonas' concept of value in being and the absence of a similar fundamental theoretical concept in Kantian ethics. Classical interpretations of Kant claim that his ethic is independent from a notion of the good and focuses solely on duties and rules founded in practical reason. This understanding holds that rules or laws function as a basis for assessment, independent of a notion of value. The problem with this approach to duty ethics is that it fails to supply a reason or justification for moral restraint. As such, the moral agent lacks a means to integrate morality into her system of ends or goals, or into a framework for moral deliberation to address complicated questions. In the arena of inheritable genetic modification, the complexity of scientific data and the profundity of the moral implications render classical duty ethics inadequate on both counts.

Contemporary philosopher Barbara Herman argues for a radical shift in the classical approach to Kant.[10] She rejects the tradition that perceives Kant's ethics as a deontology, emphasizing only duty and law detached from any notion of value as the basis of morality. Moreover, she compares Kant's argument to Aristotle's in the *Nicomachean Ethics,* contending that both begin with an inquiry into the nature of the good. Kant's *Groundwork of the Metaphysics of Morals* opens with the statement, "There is no possibility of thinking of anything at all in the world, or even out of it which can be regarded as good without qualification, except a *good will."* The classical interpretation of this declaration is that the goodness or value of the good will is not the primary motivator of ethical action, but a derivative matter. Duty or law, independent of desire, is the only motivation for pure ethical action. In contrast, Herman argues, on the basis of this text, that both Kant and Aristotle identify the subject matter of ethics as the good and the primary object of ethical inquiry as the "unconditioned good," which is the ultimate value because it imparts judgment about the goodness of actions. As such, the value of the good will guides ethical action.

Herman contends that a grounding conception of value provides Kant with the rationale necessary for moral restraint and a framework for full deliberative judgment. A notion of value offers an explanation of right and wrong characteristics of action, rendering moral standards understandable and capable of guiding a deliberative process. A concept of value that achieves this purpose necessarily includes a didactic component. In Aristotelian virtue ethics, the conception of the good plays the role of the final goal that informs the right or wrong of action. In Kantian ethics, good willing, in conformity with principles of practical rationality, represents conceptions of the good that provide the means of moral judgment and the shape of maxims. For Kant, a good maxim—not a final end—describes a valid way of acting. Thus, principles of practical rationality to some extent comprise the value content of maxims.

Herman's interpretation augments deontology in that it permits the principle of practical rationality to do the practical work that is ordinarily assigned to a concept of value. Because practical reason serves that function, Kant's imperative is able to explain the good and bad features of maxims in a way that is a relevant guide for deliberative judgment. Value content brings the Categorical Imperative and the Imperative of Responsibility into greater proximity. Jonas uses the value of the "worth in being" function in much the same way as the principle of practical rationality, because the concept of value in both imperatives provides reason for restraint and a framework for determining the right and wrong of human action.

ETHICAL LIMITS OF RESPONSIBILITY

The main thrust of Jonas' argument is to demonstrate that the former boundaries of human action and responsibility are no longer secure. The prospect of obliterating the natural through human action looms as a possibility, and matters never before considered in the purview of humankind are now relevant. Ethics is frequently called upon to reflect on theoretical future projections in light of current realities. Without the guidance of limits, ethics in this role becomes purely speculative. The more productive function for ethics is to propose limits beyond which the verity of human predictive knowledge is improbable. Jonas' philosophical basis for the identification of boundaries, in conjunction with Kant's theoretical position, assists with this challenge.

Jonas is particularly helpful in articulating the limitation of knowledge or the "uncertainty of prognostications." He begins with delineating the duty to know. Jonas acknowledges that the assurance enjoyed by the calculation of proximate consequences may not be available for future predictions. The immense complexity of the human genome, combined with the impact of changing environmental conditions, may forever preclude making confident judgments for future generations. With this assertion, Jonas emphasizes the limitations of predictive knowledge as a facet of our responsibility to know. It is at this critical point that the stipulation of boundaries of human knowledge and the limits imposed on that knowledge by human finitude are essential.

Kant's theoretical position on the "recognition of difference" serves as the philosophical grounding to delineate limits of the knowable for the purpose of genetic alteration. The focus of Kant's theoretical argument is the institution of the limits of knowledge based on the difference between what we can know and what we can merely think. In the Transcendental Analytic, a main tenet of Kant's philosophical position, Kant maintains that he proved that knowledge is limited to objects capable of being presented under the conditions of space and time.[11] Space and time are the parameters through which consciousness structures sensible experience, and understanding is the cognitive faculty that processes experiences under the conditions of time and space. That is, we are aware of a possible object of experience through our understanding. In addition to the restricted capacity of understanding, there is a more general faculty that Kant calls reason. Reason is not restricted to appearances given under the conditions of space and time, and is capable of reaching beyond the limits of possible experience. However, when reason transgresses the limits of space and time, it becomes "dialectical" and violates the very laws of logic of which it is the source.

Thus, the knowable is limited to possible objects of experience. What this means for Kant is that there is a material component to knowledge that is external and not arbitrary. As reason attempts to reach beyond the limits of understanding, fallacies or contradictions result. Consideration of objects outside of this realm is spurious. Noumenal objects are in this category, since they are nonspatial and nontemporal. As such, these objects can be thought about, but cannot be known. Everything that is knowable can also be contemplated, but the reverse is false. Everything that can be pondered cannot be known. We can only know with certainty about possible objects of experience. The first step in acknowledging limits of predictive knowledge is to recognize the difference between that which is knowable and that which is merely thinkable.

As we begin to consider the future implications of inheritable genetic modification, we approach the boundary between the knowable and the thinkable by potentially altering the very conditions of subjectivity and knowledge itself. The outcomes of inheritable genetic modification for distant progeny cannot be known with certainty. We cannot predict how deliberate changes to the gene pool will affect future individuals, physically or socially. Nor can we know the political, social, or biological environment of future generations.

Kant does not say that we should not think outside of these finite boundaries. On the contrary, he acknowledges that it is characteristic of reason to desire this complete freedom of thought. He only asserts that there is no certainty outside of these established limits. It is possible to think about the consequences that inheritable genetic modification entails, but with acknowledged limitations. The primary responsibility in the area of inheritable genetic modification is not to inhibit progress and thereby forfeit the potential therapeutic benefits, but to acknowledge human finitude and limitation, especially as it concerns modifications of the subject. An accurate reading of Kant would necessitate that genetic changes not denigrate the body, since, for Kant, life is a condition of exercising the rationality of the subject.

THE HEURISTIC OF FEAR

The discussion of boundaries kindles the consideration of the means by which ethics identifies and implements limits. Jonas contends that faith in sacred truth is capable of identifying human constraints. However, he notes that sufficiency, even in the presence of documented need, cannot activate a faith in the sacred that is departed and disparaged. Consequently, he offers the alternative of fear to replace the awe of the sacred as a means of identifying a boundary.

Fear

In light of the modern denuding of the sacred, Jonas proposes fear as the prevailing means of regulating the power to act. For Jonas, the purpose of ethics is "for the ordering of actions and for regulating the power to act." Obviously, religion in regress cannot accomplish the task. In the vacuum that remains, "Fear can do the job—fear which is so often the best substitute for genuine virtue or wisdom."[12] The fear proposed by Jonas results from perceived evil.

He contends that while we sometimes find it difficult to define the good, it is usually possible to identify evil when it appears.

While Jonas is very clear on this point, it is possible to argue that the designation of evil is equally problematic. Yet, he forcefully asserts that it may be through the experience of the evil that the good is recognized. Jonas argues that it is fear of the "anticipated distortion" of the human species that initiates a normative concept of humankind. It is the "threat to the image of man" that brings one to cringe before the threat of remodeling the genome. The knowledge of what to preserve comes directly from the ability to recognize what needs to be avoided. When fear is not generated by evil already present, Jonas maintains that it is our duty to seek out fear by an effort of reason.[13]

Jonas believes that the heuristic of fear, rather than the perception of good, should assume the forefront in ethics. For Jonas, the past concentration on the Socratic premise of the good is a conundrum. He believes that, in reality, our fears are a preferable guide to morality. Arguably, it is sometimes easier to identify that which is abhorrent as opposed to that which is truly beneficial in human genetics. However, to limit ethics to the assessment of the abhorrent is to leave unattended the vast majority of cases that are neither clearly abhorrent nor acceptable but require subtle ethical distinctions. For example, the endeavor to diminish human intelligence with the purpose of creating a race of willing slaves is morally reprehensible. On the other hand, the enhancement of memory is neither obviously abhorrent nor clearly acceptable since it entails both potential benefits as well as possible injustices. It is difficult to assess the balance of good and harm in these more complex cases, operating with fear as a guide to morality.

Moreover, the method of societal implementation that Jonas recommends for the heuristic of fear is especially problematic. Jonas sees Marxist ideology as the only possible source of an ethic that proposes norms for action that extend into the future. The Marxism that Jonas poses would be almost unrecognizable when compared to the twentieth-century version, except for the organizational principle of a classless society. The divergence lies in Jonas' reinterpretation of the role of Marxism from the deliverer of utopia to the "preventer of disaster." His version of the classless society would function to ensure humankind's survival rather than to fulfill the utopian dream. In this role, fear of doom rather than the promise of salvation grounds a Marxist society.

The heuristic of fear is a concept that contributes to the discussion, but one not endorsed by this author as applicable to the governance of genetic technology. To sanction or prohibit action based on fear has the potential to immobilize that which is essential to genetic research. Freedom, not suppression, nurtures the goods internal to the practice of human genetics. For this reason, the heuristic of fear does not play a central role in the development of the Inheritable Modifications Matrix. Instead, I adopt an ethical perspective consisting of pertinent virtues, duties, and principles grounded in human dignity to address the challenges of genetic power.

NOTES

1. Hans Jonas, "Ethics and Biogenetic Art," *Social Research*, no. 52 (Autumn 1985): 491–504.
2. Hans Jonas, *Philosophical Essays* (Englewood Cliffs, N.J.: Prentice-Hall, 1974).
3. Hans Jonas, *The Imperative of Responsibility: In Search of an Ethics for the Technological Age* (Chicago: University of Chicago Press, 1984).
4. Jonas, *Philosophical.*
5. Jonas, *The Imperative.*
6. Jonas, *The Imperative.*
7. Immanuel Kant, *Groundwork of the Metaphysics of Morals,* trans. and analyzed by H. J. Paton (New York: Harper & Row, 1956).
8. Jonas, *Philosophical.*
9. Kant, *Groundwork.*
10. Barbara Herman, *The Practice of Moral Judgment* (Cambridge, Mass.: Harvard University Press, 1993).
11. Immanuel Kant, *The Critique of Pure Reason,* trans. Norman Kemp Smith (New York: St. Martin's Press, 1965).
12. Jonas, *The Imperative.*
13. Jonas, *The Imperative.*

7

The Circumstances: What Is Her Context?

How can I coexist with him and still leave his otherness intact?

—Emmanuel Levinas, *Totality and Infinity*

As analysis of the moral event progresses, the circumstances become the reference for identifying principles to include in the Inheritable Modifications Matrix (IMM). Circumstances refer to the cultural context, or setting, of the moral event. They are those conditions that define a person's particular traditions, habits, beliefs, genetic inheritance, or psychosocial state. Levinas's ethic of the Other was chosen to address the context in the ethical event, since his focus on the multiplicity in being represents the ethic of particularity. Particularity is an approach to ethics that reflects on the values and traditions of different cultural and ideological groups in a respectful and deliberate manner.[1] It counters ethical approaches like that of Jonas, which attempt to surmount cultural differences, with universal principles that apply to all persons in most situations.

In order to orient Levinas with respect to Pellegrino and Jonas, it helps to reflect briefly on their similarities and differences. Most important to the task at hand, all three share a fundamental commitment to the inherent value and dignity of the human person. Pellegrino grounds human dignity in a Catholic incarnational theological view, which attributes intrinsic value to human beings as created in the image of God. Jonas derives dignity from an ontological analysis that grounds human value in the property of purposiveness. In comparison, Levinas's derivation of dignity comes from a phenomenological focus on the particular experience of the individual Other. The common theme in these three diverse ethical approaches is the emphasis on the dignity of every human being and the corresponding injunction against the destruction of life.

THE LANGUAGE OF THE ETHIC OF THE OTHER

Levinas's philosophy is postrational, and the dramatic quality of his language reflects its opposition to the intellectual. The vocabulary of the Other describes the subjectivity experienced in the human situation and complements Levinas's construal of ethics as a relationship with complete difference. The purpose of the brief discussion that follows is to familiarize the reader with Levinas's vocabulary and provide the basis for the identification of his principles for inclusion in the IMM. The terms used throughout this chapter are the basic language of Levinas's ethic of the Other.

For Levinas, the Other is the other person who is completely separate from the self. The Other is not another self who is essentially similar but possessing her own set of particularities, nor is the Other an object categorized under standard classifications and assumed neatly into an organized worldview. Instead, she is other than the self, a stranger, and the self cannot know what this stranger conceals.[2] Levinas labels the self who desires to thematize the Other as the "Same." The Same is the subject who knows and possesses itself; it is the autonomous self. The Same represents a "for itself" mode of existence that accepts its own egoism as innate as the will to live.

Levinas grounds the ethic of the Other in the presence of transcendence and infinity within human experience. The infinity present in the gaze of the Other is a sign of transcendence. Transcendence in this context does not imply any dogmatic content; it reflects the dimension of height from which the Other encounters the Same. Transcendence signifies the dimension that is beyond experience and cannot be comprehended within ordinary limits. Thus, the usage of transcendence in Levinas signifies the metaphysical component of the Other. Transcendence that presents itself in the face of the Other is the articulation of alterity and dignity. Levinas uses the term "alterity" to denote remoteness and the dimension of height from which the Other approaches the Same. Alterity does not represent the counterpart of the identity of the self, but the very content of difference, of otherness.

The notion of the infinite in the face of the Other expresses the radical asymmetry of the self's relationship with the Other. Infinity designates the presence of that which always overwhelms the thought that thinks it. The mind determines what it discovers by itself, but the presence of the Other always exceeds this conceptualization. The notion of infinity expresses the significance of difference and separation in the ethical relationship that is primary for Levinas. From this foundation, he asks: "How can I coexist with him and still leave his otherness intact?"[3]

The purpose of the use of Levinas to consider the circumstances of the individual human person is threefold. First, the intent is to discern principles that represent the context of the Other for incorporation into the IMM. The principles of *discourse* and *difference* emerge from this examination of particularistic ethics. Second, Levinas's work reinforces the designation of the principle of *responsibility* as a component of the matrix, as discussed in chapter 6. Levinas calls for limitless responsibility as the core of the ethical relationship with the Other. Third, this inquiry sheds important insight on the ethical limits beyond which the pursuit of inheritable modification becomes unethical. The emphasis on the significance of the particular circumstances of the Other provides a unique perspective on all three endeavors.

THE ETHICAL RELATIONSHIP WITH THE OTHER

Levinas defines ethics as the originary relation to the Other that is prior to any ontology or science of norms. "The ethical relation is not grafted on to an antecedent relationship of cognition, it is a foundation and not a superstructure."[4] For Levinas, ethics is the face-to-face relationship with the Other that eventuates in language. This definition of ethics culminates in the inclusion of Levinas's principle of Discourse, or informed consent, as a component of the matrix.

For Levinas, ethics is spiritual. In the essay "God and the Idea of Infinity," Levinas presents ethics as "otherwise than being," preceding the autonomy of the self.[5] His intent is to lead beyond theory to a relationship that does not reduce the Other to the Same as does ontology. Levinas criticizes ontology's attempt to "totalize" the Other through the comprehension of Being. Rather, in the presence of the Other, the Same confronts a defenseless power that questions the imperialism of the ego. "The strangeness of the Other, his irreducibility to the I, to my thoughts and my possessions, is precisely accomplished as a calling into question of my spontaneity, as ethics."[6] The welcoming of the other in her dissimilarity, and the acknowledgment of the presence of infinity in her face, is the basis of the ethical relationship. The dimension of separation or loftiness in the gaze of the Other is the source from which the privilege of the Other ensues.

The infinity of the Other is the source of superiority that simultaneously brings the Same into existence and contests the freedom of the Same. The concept of superiority recognizes the asymmetry in the ethical relationship with the Other, and the presence of infinity calls the ego of the Same into question. Freedom resides in the dominance of the Same and the negation of the Other. Before the face of the Other, freedom retreats; yet, the Other does not negate the freedom of the Same, but "founds it and justifies it." The Other, whose eyes express the ethical mandate against killing, impacts the freedom of the Same. If murder of the Other is ethically inconceivable, the power of the Same ceases. As such, freedom is reframed.

"It is in the laying down by the ego of its sovereignty . . . that we find ethics."[7] For Levinas, ethics is not merely a branch of philosophy, but is the "first philosophy." The ethical relationship is not dependent upon a prior relationship of cognition; instead, it is primal. To set ethics apart from cognition is not to reduce it to mere sentiment. The concept of infinity is the crucial notion in clarifying Levinas's ethics. Infinity signifies that attribute of the Other where "being overflows the idea, in which the other overflows the same."[8] The relationship with that which truly exceeds thought is termed experience, a relationship with exteriority or difference. The separation from the Same, which establishes the Other as truly other, is exteriority. It is in this experience with difference that ethics precedes ontology and is separate from mere sentiment. As such, "The desire for infinity does not have the sentimental complacency of love, but the rigor of moral exigency."[9]

Face

The concept of the face is the central focus of Levinas's ethics. "The way in which the other presents himself, exceeding *the idea of the other in me,* we here name face."[10] As such, the face is not manifested merely in its qualities, but expresses itself and offers an individual context that inundates all prior cognition and generalization. In its reality the face is nude, without any cultural ornament. The nudity of the face reveals an "ultimate strangeness" that transcends the ordinary. This foreign presence eludes the grasp and disturbs the accepted order of events. The infinity present in the face of the Other defines a separation that cannot be defiled.

The face of the Other reveals the separation that resists the power of the Same to grasp or comprehend. In the denial of comprehension exists the possibility of the ethical. The relationship is ethical because it is not the simple reverse of identity or the addition of common concepts; it is a relationship with difference. For Levinas, the face-to-face experience is the essence of the ethical relationship. The straightforwardness of this position precludes the Same from negating the Other. The epiphany of the face opens up a discourse based on an ethical obligation to respond practically to the Other's appeal. The bond between discourse and the obligation of inordinate responsibility is the ethical condition, a relationship with infinity expressed in language.

"Saying"

Levinas uses the term "saying" to denote proximate discourse with the Other. The principle of Discourse in the matrix ensues from the Levinasian mandate to pursue the ethical relationship in language. Saying represents the opposite of presumed understanding of what the Other would say. The requisite is that the Same approach the Other in speech and acknowledge the presence of complete difference. The recognition of the ultimate demand signified by the face of the Other initiates the ethical relationship.

The function of language in this relationship is to invoke the Other. In the process of conversation, the Other, despite the relation with the Same, remains transcendent. Language spans the distance between the entities without eliminating it. Language acknowledges and invokes an absolutely separate person. It is the asymmetry in the relationship with the Other that renders dialogue "utterly indispensable and hugely difficult."[11] In the process of discourse, one does not question oneself regarding the Other; instead, one questions the Other. Thus, this process is not reducible to the thought of the Same. The Other communicates ideas to inform the Same, who in return can assent or call the information into question.

LEVINASIAN PRINCIPLES

Clearly, to formulate a set of principles and rules to guide the course of inheritable genetic modification is not ethics in the fundamental sense, as defined by Levinas. While Levinas speaks in terms of limitless obligations, he consistently avoids succumbing to a universal ethic that would impose duties upon all rational beings. This is because principles are universal standards for moral action that tend to reduce the Other to a duplicate of the Same. The formulation of policy typically obscures the Other under the rubric of generalities and eliminates the opportunity to listen to the saying of the Other. The perpetrators of the principles or policies impute their own moral beliefs to the Other, obliterating the attempt to experience or respond to difference. The problem becomes exponentially more serious with the ability to generalize the future Other. To relegate the future Other to a categorization of the perfect human is to reduce her to an abstraction of personal choice, a preordained concept. For example, the genetically engineered human being falls under a schema of cognition that the Same brings to bear upon its creation, and ultimately the Other becomes an object or possession.

The application of Levinas's mandate of face-to-face discourse to inheritable genetic modification presents challenges. A tension clearly exists between the requirement of proximate discourse (or saying) of the Other, and the eventual need for principles that are capable of guiding the power of biotechnology. Levinas would necessarily mistrust all principles because they are overarching standards. He repeatedly argues that ethics is a fundamental relation with the Other, contingent upon discourse, and that the sayings of the Other are the only legitimate starting point for framing a notion of the good of the Other. Following from that, principles are valid only if revisiting of the Other's sayings is a provision included in their creation of policy. While the formulation of principles is not the work of ethics in the Levinasian sense, he doesn't exclude the attempt to express general moral standards with the above stipulations.

Author and professor of philosophy Jere Surber suggests that any attempt at a Levinasian principle should include the following criteria.[12] First, such a principle should emphasize the obligation to react to the desires of the Other as the Other would have the Same respond. This requirement in the formulation of principles reflects a change in focus from the self to the other, in which the values of the Other take precedence over those of the Same. Second, a Levinasian principle would necessitate that the Same never predetermine what the Other would request: responsive listening to the discourse of the Other is paramount. The only justifiable means of ethical reflection is the experience of the face-to-face relation with the Other, consummating in language. The third requirement for a Levinasian principle focuses on the prohibition against murder. The foundational mandate is "Thou shall not kill"; any qualifiers to this mandate would negate the continuance of the ethical relation. For example, the principle could never require the death of the Other or of the Same or any type of self-sacrifice, unless the life of the Other were in danger. The human relationship must be preserved in order for any principle to be ethical.

Therefore, the adherence to the saying of the Other, or the principle of discourse, grounds the ethical relationship. In accordance with Levinas's ethic, discourse is a component of the matrix. If the process of determining the morality or immorality of inheritable genetic modification is to be ethical, albeit not in the most fundamental sense according to Levinas, all principles or policies must originate in prior listening to the conversation of the Other. Levinas also cautions that there must always be the "possibility of unsaying" in the ethical relation. As such, it is important that proximate discourse involves a rhythm that sanctions the saying of the Other to be periodically reformulated. This reinforces the continuation of conversation with the Other, who is the potential subject of any public policy, procedure, or research involving human genetic modification. Another way of expressing this mandate is the familiar principle of informed consent. Obviously, this mandate will almost always preclude inheritable genetic modifications that alter the genome of future individuals, since we cannot obtain informed consent of another that does not yet exist. I will elaborate on those procedures that are commonly exempt from informed consent in chapter 9.

RESPONSIBILITY FOR THE OTHER

"Ethics is language, that is, responsibility."[13] The requisite to listen and respond to the saying of the Other places the ethical obligation of responsibility on the Same. Responsibility before the face of the Other is a summons that reinforces the identification of the duty of responsibility as a component of the matrix in chapter 6. In the context of the ethic of the Other, responsibility takes on a different meaning. It entails viewing one's actions and omissions from the perspective of the Other's desires. Before the face of the Other, the Same is "inescapably responsible." The Same becomes responsible for the misfortune and suffering of the Other, for everyone, and even for that which the Same has not prescribed. The responsibility of the Same for the Other does not include a reciprocal responsibility, given the asymmetry of the ethical relationship. Yet, this does not preclude the possibility of respect and fair treatment for the Same. This is made possible by Levinas's concept of the third party, as was discussed in chapter 5, in whose presence the Same becomes another Other.

Consummate responsibility dissipates the egoism of the Same, but it does not destroy one's uniqueness. Levinas protects the uniqueness of the Same in that no one can respond to the Other for someone else. "Inescapable responsibility" obliges each person as a unique entity chosen to respond; it cannot be designated to a proxy. No one escapes the summons in the face of the Other, without negligence in the ethical relation. According to Levinas, there are also no limits to the summons of responsibility.

Levinas's emphasis on responsibility as the fundamental of the ethical relationship contrasts with that of Jonas. For Jonas, responsibility is the mutual relation of power and knowledge. He asserts that power increases as the magnitude and scope of responsibility changes, with the effect that power generates the content of the ethical obligation. In other words, duty springs from the power to act. For Jonas, fear for the universal continuance of the human race is the impetus for the imperative of responsibility.[14] Jonas expresses the affirmation of being abstractly or rationally. This is a form of the all-encompassing or totalizing thought based on ontological priority that Levinas denounces.

The philosophical divergence in the grounding of responsibility may seem incompatible at first glance. However, in the juxtaposition of these two dissimilar approaches to ethics, and particularly to the duty of responsibility, some analogous philosophical concerns resound. Both Levinas and Jonas esteem the ultimate value of all forms of human life and ground responsibility in the obligation to protect the continuation of life, albeit from the different viewpoints of metaphysics and ontology, respectively. Additionally, both philosophers place the responsibility for the preservation of life on the ethical agent.

THE PLURALITY OF THE OTHER: DIFFERENCE

For Levinas, plurality denotes complete otherness. Plurality is not an addition of Sames, but a multiplicity in which the Same relates to the Other. It is not a numerical multiplicity, but one in which individuals retain their diversity in relation to each other. The significance of the term "plurality" in Levinas is quite different from that in common usage. The prevalent usage of "plurality" connotes "the quality of being plural . . . the majority." This and other contemporary connotations are inconsistent with the compelling symbolism of plurality in Levinas. To avoid misunderstanding, I will utilize his term "difference" to represent the Levinasian focus on complete otherness in the matrix. The substitution of the term "difference" represents Levinas faithfully, while avoiding the confusion that the term "plurality" generates.

Difference, in Levinas's writing, "implies a radical alterity of the Other, whom I do not simply *conceive* by relation to myself, but *confront* out of my egoism. The alterity of the Other is in him and is not relative to me; it *reveals* itself."[15] The pluralistic point of view mandates that ethical judgments be made "with reference to a standard of perfection that is radically other and transcendent." Those dissatisfied with the status quo espouse pluralism; Levinas calls such persons "infinitizers." Those who champion this ethic seek a higher quality of life for all individuals based upon a societal perspective that is creative and esteems difference. The standard of difference and transcendence held by infinitizers welcomes the metaphysical value of human beings as subjects rather than as objects. This is a radical departure from "totalizers," who devise standards and laws from sameness. The intent of totalizers is to enforce order and systems that result in power and control. Under their systems, only a preconceived theoretical construction of Being qualifies, and only that which can be objectified by the mind is real. Objective thinking is the priority, and all that is emotional or experiential is discounted as subjective.

Obviously, very different responses to the challenges of inheritable genetic modification issue from proponents of the antithetical ethical viewpoints of totality and infinity. As we scan human history, the atrocities attempted through genetic means are often the repercussions of totalizing thought that attempts to reduce the Other to a copy of the Same in the name of societal advancement. If such atrocities are not to recur, the mandate of the commitment to difference is essential. Beings who are truly Other cannot be designed or owned by others.

The relationship with complete difference is not one of possession, but of separation. Even offspring are actually strangers in the homes of their parents because every child bears a genetic composition that is unique and overwhelms any preconceived idea that parents have. The ethic of the Other esteems the transcendence present in the face of a child who is not an object of technology but a subject unto herself. It also embraces the obligation to acknowledge difference as an essential component of the ethical relationship.

THE LIMITS OF THE ETHICAL RELATIONSHIP WITH THE OTHER

Part of the task of this book is to explore the issue of limits that plagues the utilization of inheritable genetic modification. "How far should we go?" and "When should we stop?" are questions frequently asked. Levinas's work offers valuable direction in understanding the concept of limits as it applies to the ethical relationship. From this starting point, in conjunction with Kant's interpretation of limits, some enlightening applications emerge.

In an attempt to clarify the discussion of ethical limits in Levinas, an analysis of otherness in Kant offers some assistance. This method identifies some of Levinas's themes in a way that facilitates practical application. A common critique of Levinas is that his "excessive" language precludes comprehension. In response, Surber has attempted to temper the alleged excessiveness of Levinasian ethics with the theory of Kant.[16] This is not to imply that Surber's inquiry resolves all the crucial differences between Kant and Levinas; it is only a means to utilize a different philosophical expression to interpret the significance of limits. Without claiming to summarize Surber's very complex analysis, I will recount the Kantian concepts of *reflexion* and *nichts* as a path to articulate the limitations of knowledge, discourse, and experience posed by the presence of the Other.

Reflexion

Surber chose the Kantian notion of *reflexion* to articulate the difference that is crucial to the ethical relation that Levinas describes. *Reflexion* is discussed in an appendix to the Transcendental Analytic in the *Critique of Pure Reason*.[17] It is the contemplation of difference from the acknowledgment of difference. It signifies the process that expresses the difference between the sensible and the conceptual, without the goal of establishing a higher identity to encompass both. *Reflexion* does not involve the transcendental dialectic and the tendency to free reason from all restrictions, resulting in fallacy; its importance is its acknowledgment of irreducible difference without the necessity of an overarching concept. Surber compares quantitative and qualitative *reflexion* to Levinas's concept of infinity and to the prohibition of murder, respectively.

Quantitative *reflexion* asserts that any general concept of human nature to which all other beings are relegated is perpetually inadequate since it ignores the insurmountable differences between the conceptual and the sensible. In Levinas's terms, a generalized concept of human essence cannot encompass the infinity of the Other, since the infinity of the Other transcends categorization. On the basis of this comparison, Surber correlates Kant's quantitative *reflexion* to Levinas's "beyond being," which indicates that ethics is prior to any attempt to comprehend Being. Quantitative *reflexion* focuses upon the Other's conception of her own good, which is open and receptive to further sayings of the Other. Applied to inheritable genetic modification, a standardized concept of genetic perfection can never capture the depth and breadth of the human spirit or serve as a template for life. Any procedure, research, or policy endorsing the production of a human being according to a preconceived model transgresses ethical limits. It denies the possibility of the conception of one's own good and is unresponsive to the sayings of the Other.

To further clarify the notion of limits, Surber compares qualitative *reflexion* to Levinas's imperative "Thou shall not kill." Qualitative *reflexion* refers to an object reduced to nothingness, and explains Levinas's assertion that the primary ethical imperative is "You shall not kill." To murder the Other violates the very foundation of the moral order, for it negates the relationship upon which ethics depends. Only the presence of the Other prompts the Same to respond to the duty of the moral law. Murder is the absolute destruction of difference; it is to see the imperative in the eyes of the Other and respond with a violence that nullifies all ethical dimensions. Civil and religious laws almost universally forbid the actual act of murder. Moreover, qualitative *reflexion* implies that the murder of the Other can occur as a matter of degree. First, one simply ignores the face of the other, then discounts her discourse, next deprives her of the right of discourse, and finally deems the Other as ethically irrelevant and unworthy of existence. The common practice of "murdering" the Other by neglect, abandonment, degradation, or derision is the ethical analogue to physical murder. Every action that deems the Other undeserving of her individuality is a murderous gesture that transgresses the limit.

Nichts

The second concept is Kant's account of *nichts,* or "nothing." It is found in an obscure appendix to "The Amphiboly of the Concepts of Reflection" in the *Critique of Pure Reason.*[18] *Nichts* is a way of clarifying the limits of our knowledge: It expresses a means of reflecting on the Other that neither forsakes her to the realm of the inconceivable, nor allows her to be consumed in the identity of the self. Rather, Kant supplies us with a means of orientation within an asymmetrical relationship with absolute difference. It is a "logic of the Other," which does not deem him totally unknowable but acknowledges the asymmetry between the Other and the self. *Nichts* does not reject the possibility of placing oneself in an asymmetrical relation to the Other, but merely forces us to acknowledge the human limitation in the presence of difference.

Kant articulated *nichts* with a fourfold schema. I will emphasize the two elements of this schema that are especially germane to the establishment of limits in the presence of the Other. First, the notion of quantitative *nichts* indicates awareness of certain concepts that we can think will be inadequate for their intentional objects. It is an awareness of the otherness of what is Other. Thus, the concept is not empty but simply inadequate for its proposed object. It marks the asymmetry of difference. The proposed object of the concept always exceeds the concept used to comprehend it. The application of quantitative *nichts* to inheritable genetic modification points out that one cannot conceptualize fully the desires, dispositions, or aims of the Other, especially the future Other. To attempt to do so transgresses the ethical limit, since the desires of the Other always exceed conceptualization. Thus, the limit is the acknowledgment that one can never know definitively, completely, and exhaustively the desires of the Other and, consequently, cannot ethically attempt to program her genome.

It is also possible for the opposite to occur: a theoretical concept, instead of being inadequate for its object, can overpower it. This is qualitative *nichts,* or the place in reflection where one recognizes that the concept vanquishes its object, diminishing it to the status of nothing.

This is not to say that there is the absence of an object, only that the attempt to comprehend an object reduces it to unimportance. This notion is manifest in human genetics as the reduction of the Other to an analysis of her DNA. Consider the case of an adolescent with cystic fibrosis who is a good student and a tremendous athlete. He knows the implications of the genes he carries, but resolves to make the most of his talents every day that he can and be an example to others suffering from this disorder. His genetic sequence reveals the presence of genes that are fatal, but it cannot represent the courage and determination of this young man or describe the significance of his life experience.

Kant's notions of *reflexion* and *nichts* illustrate the discrepancy between the conceptual and the sensible, and express the limits of the ability to fathom the difference. This analysis explores the correlation between Kant and Levinas in an endeavor to articulate ethical limits in the presence of the Other. The temporal Other can never be relegated to a conceptual "object of knowledge." To do so is to transgress the limit prohibiting annihilation. Any attempt to reduce the Other to insignificance is to commit murder in the truest, even if not the physical, sense of the term. In addition, the notion of *nichts* is a means of articulating the limitation of human knowledge in a concrete fashion. *Nichts* illustrates the asymmetry between concept and object in the presence of difference. Genetic modification transgresses ethical limits when it claims the ability to determine and control the mystery of life.

Levinas barrages this hubris. He provides a way to think about the Other that mandates wonder, not domination, in our relationship with her. He demands that we contemplate the face of the Other that expresses vulnerability and asks not to be injured. He obliges us to listen to her silent injunction that utters, "Thou shall not kill." The ethic of the Other is fertile ground for contemplating the challenges of inheritable genetic modification.

NOTES

1. For a discussion of the ethic of particularity, see Daniel Callahan, "Universalism and Particularism," *Hastings Center Report* 30, no. 1 (January–February 2000): 37–44.
2. In an attempt to be faithful to Levinas's terminology, the personal Other, which is in the second person, is indicated by "Other" and the general other, which is in the third person, is represented by "other."

3. Emmanuel Levinas, *Totality and Infinity,* trans. Alphonso Lingis (Pittsburgh: Duquesne University Press, 1969).
4. Emmanuel Levinas, *Collected Philosophical Papers,* trans. Alphonso Lingis (Boston, Mass.: Martinus Nijhoff Publishers, 1987).
5. Levinas, *Collected Philosophical.*
6. Levinas, *Totality.*
7. Emmanuel Levinas, "Ethics as First Philosophy," in *The Levinas Reader,* ed. Sean Hand (Cambridge, Mass.: Blackwell, 1993), 11–28.
8. Levinas, *Collected Philosophical.*
9. Levinas, *Collected Philosophical.*
10. Levinas, *Totality.*
11. David Morris, "How to Speak Postmodern: Medicine, Illness, and Cultural Change," *Hastings Center Report* 30, no. 6 (November–December 2000): 7–16.
12. Jere Surber, *Ethical Principles: Levinas and the Tradition* (Denver: University of Denver, 1996).
13. Levinas, *Collected Philosophical.*
14. "Fear for," as used by Levinas, denotes a concern for the preservation of the life of the other. This is different from "fear of," as used by Jonas, to indicate an expectation of retaliation if certain injunctions are transgressed.
15. Levinas, *Totality.*
16. Jere Surber, "Kant, Levinas, and the Thought of the Other," *Philosophy Today* (Fall 1994): 104–116.
17. Immanuel Kant, *The Critique of Pure Reason,* trans. Norman Kemp Smith (New York: St. Martin's Press, 1965).
18. Kant, *Critique.*

8

The Consequences: How Can We Be Happy?

If we do not know where we are going, and perhaps can not know, can we agree on where we should *not* be going?

—Daniel Callahan, *Medicine Unbound*

As we launch into investigating the final aspect of the moral event, the consequences, a summary of previous conclusions is useful. Thus far, the analysis includes three components of the moral event: the agent, the act, and the circumstances. Three different ethical theories contributed to the identification of virtues, duties, and principles pertinent to these three facets of moral decision-making. The analysis of the agent focused on virtue ethics and the cultivation of virtues necessary to ensure the moral use of inheritable genetic modification. The pertinent virtues identified by that inquiry are *integrity, justice,* and *wisdom*. The examination of the ethical act explored human action in light of duty or obligation. The duty required is *responsibility* for the continuation of human life. The study of the circumstances of the moral event centered on the principles that underscore the significance of individual particularities; the principles are *discourse* and *valuing difference*.

UTILITARIAN ETHICS

As stated, this chapter addresses the final aspect of the moral event, the consequences. Utilitarianism is the most prevalent consequence-based theory. The origins of utilitarianism came from the writings of Jeremy Bentham and John Stuart Mill, philosophers whose work stretched from the mid-eighteenth to the mid-nineteenth centuries.[1] Utilitarianism recognizes only one principle as the basic criterion of ethics: utility, which Mill describes in terms of happiness or pleasure. The principle of utility requires the maximization of social utility, as defined as the arithmetic mean of the happiness level of all members of society: the greatest good for the greatest number of people. In addition, utility places substantial importance on the value of individual choice in ethics. Mill claims that the very measure of an individual is the extent to which one makes life choices free of pressure.

Utilitarian consequentialism calculates benefits and harms as a quantitative way of determining the morality of an action. The calculation requires that ethical actions result in overall pleasure, or produce the least amount of harm if good results cannot be achieved. Mill proposes a universal ethical hedonism when he contends that the greatest happiness for the greatest number be the goal and standard of conduct:

> Actions are right in proportion as they tend to promote happiness, wrong as they tend to produce the reverse of happiness. By happiness is intended pleasure, and the absence of pain; by unhappiness pain, and the privation of pleasure.[2]

Mill contests Bentham's quantitative hedonism that ranks all pleasures as equal, and contends that the discrepancies among pleasures are subject to important qualitative dissimilarities. He espouses a qualitative hedonism that ranks higher pleasures, such as reading ethics, above what some may deem lower pleasures, such as consuming a good meal. This creates a problem for utilitarians, because it raises the question of who shall judge the qualities of the various pleasures. Mill states that we ought to defer such judgments to the "experts," those knowledgeable individuals who have experienced the entire spectrum of pleasures and can differentiate among them. This statement generates the further question of whether such competent experts exist and if their judgments can be applied universally. Mill claims that those who are equally acquainted with all types of pleasure prefer those that employ the higher faculties. Is it accurate to say that the intellectual pleasure of discovering a gene always ranks above the enjoyment of watching one's favorite football team? Mill leaves the final ranking of pleasures open to discretion, but his qualitative approach does avert gross misconstruals of what counts as pleasure and pain.

The evaluation of actions as right or wrong based on the calculation of happiness permits utility to deal with ethical quandaries in terms of empirical facts without referring to a restrictive standard of human nature or moral obligation. In addition, it is indisputable that good should be maximized and harm minimized. There are many examples in genetic medicine that demonstrate the practicality of utility. The perinatal test for phenylketonuria (PKU) is one such example. PKU is an autosomal recessive genetic disorder characterized, if untreated, by microencephaly and profound mental retardation. A strict diet, if begun in early infancy, can prevent significant mental retardation. The diet is expensive and undesirable, especially when the child acquires opinions about what he or she chooses to eat. However, the value of normal cognitive function far outweighs the disvalue of an undesirable diet or the ravages of profound mental retardation. On the basis of the calculation of the benefit:harm ratio in the treatment of PKU, there is compulsory screening for the disorder at birth in many states. On the basis of numerous such examples, utility describes theoretically what is already implicit in everyday practice.

ACT AND RULE UTILITARIANS

The prevention or minimization of pain is the stated goal of molecular genetic research. Pain is the harm caused by intractable genetic disorders, scientific errors, or unforeseen consequences. Pleasure or benefit in genetics equates health and well-being. If inheritable genetic modification becomes a standard treatment for suitable genetic disorders, should the benefit:harm ratio expand to collectively include present and future persons? If so, how? A collective good is a larger, all-encompassing benefit that transcends and includes individual benefit. The collective or overall good is not a commonly espoused notion in contemporary culture. Most Americans prefer autonomy and the pursuit of personal independence to an altruistic consideration of the communal or societal good.[3] Many ethical disagreements among thoughtful people result from differing viewpoints about the future consequences of alternative policies on the individual versus collective levels.

Two schools of utilitarian ethics mirror these differences: act utilitarians and rule utilitarians. While both affirm the maximization of social utility as the absolute standard of ethics, they differ over whether the principle pertains to specific acts or to general rules. Act utilitarians estimate good or bad consequences likely to result from a particular act in a given circumstance, assuming that the task is to maximize social utility. They accept pertinent moral rules as useful, but dispensable if the rule does not maximize utility in the context at hand. In contrast, rule utilitarians apply a general utility-based rule to a type of action in a common set of circumstances that all must follow. For rule utilitarians, conformity to the rule determines whether an act is right or wrong; the general rule holds, even if it does not maximize utility in the given set of circumstances.

Consider how an act utilitarian versus a rule utilitarian might approach the case of John, a young man who tested positive for familial colon cancer, placing him at high risk of developing cancer by forty years of age. His mother died of colon cancer, and he has an adult sister and brother who may have inherited the mutation. He does not plan to share his diagnosis with them because they both suffer from bouts of depression, and he fears that the news will worsen the condition. Genetic information, like all other medical information, is protected by the legal and ethical principle of privacy. In this instance, John intends to exercise that right.

However, privacy is not absolute. In cases involving a treatable or preventable fatal disease, the accepted practice is to warn at-risk parties.[4] In John's case, disclosure maximizes utility on the collective level, since timely testing of his at-risk family members minimizes or avoids their serious harm and suffering. Inherited colon cancer is a fatal disorder for which there is a means of prevention; therefore, a rule utilitarian could require disclosure of risk to John's brother and sister, despite the possibility of triggering depression in them. This is based on the accepted standard that the aggregate benefit of avoiding fatal disease is greater than the harm of emotional upheaval.

Conversely, an act utilitarian could argue that disclosure of a fatal but treatable disorder does not maximize the individual's good if it results in stress, anxiety, or other harms. In John's situation, act utilitarians may treat the responsibility to disclose preventable disease as an expendable guideline even though there is an opportunity for significant collective value. In addition, act utilitarians further justify their position by stating that traditional moral rules cannot always speak to new technological developments.

UTILITY IN PRACTICE

Most of us as everyday ethicists rely heavily on the principle of utility when faced with choices that involve right or wrong. This is especially true in bioethics. Utility brings important strengths along with noteworthy limitations to the ethical assessment of inheritable genetic modifications.

Assets

Utility is ostensibly compatible with scientific thought, given its focus on the empirical consequences of action that result in harm or benefit. Primarily, an assessment of the harm versus the benefits of any genetic modification causes the geneticist to reflect on the real-world risks of altering the genome. Utilitarian considerations underscore the extreme complexity of DNA and the uncertainties of inheritable genetic moderation to produce overall benefit. When approached thoughtfully, utilitarian reasoning makes a case for humility in a situation where future individuals may encounter potential harm.[5]

Second, utility responds to the moral crisis in bioethics and contemporary culture in general. In the absence of an acceptable anchor for moral deliberation based in religion or human nature, utility renders ethical questions empirically analyzable. Individuals can calculate verifiable quantities instead of relying upon metaphysical distinctions, avoiding a stalemate caused by differing worldviews. Third, utility plays an important role in advising public policy formation. The principle of utility requires that everyone's interests count and that all affected parties carry equal weight in ethical considerations. According to the principle of utility, all humans are subjects of practical reasoning and all command respect. Finally, while utilitarianism stresses consequences, it also encompasses the principle of beneficence as a concept of value. Utilitarian determinations of morality involve promoting the greatest good. A theory of utility balanced by the principle of beneficence responds to some of the criticisms cited in the following sections of this chapter. For these reasons, this ethical theory is indispensable in practical ways to the assessment of inheritable genetic modification.

Limitations

The most prevalent critique of utilitarianism is that, according to the principle of utility, a sufficiently propitious outcome of an action could justify an immoral means. Suppose that the only means to maximize utility is to perform an act that is ethically unacceptable by common standards of morality.

The most prevalent comment on utilitarianism is that, according to the principle of utility, a sufficiently propitious outcome could justify an unethical means to that end. Let's use the widely debated issue of therapeutic cloning to illustrate how a focus on consequences can undervalue other important considerations. The goal of therapeutic cloning is to obtain genetically matched embryonic stem cells that can be steered with biochemicals to produce compatible tissues and organs to replace or regenerate diseased ones. The science of therapeutic cloning involves the generation of a human embryo that has the same genetic signature as the patient. The embryo is allowed to develop for about five days and then is destroyed to extract embryonic stem cells. Stem cells produced through therapeutic cloning have an advantage over those harvested from embryos resulting from IVF or aborted fetuses in that they are immunologically matched to the patient and thus avoid the need for life-long anti-rejection therapy.

The widespread utilitarian argument cited in favor of therapeutic cloning focuses on the benefits versus the harms of the procedure. Therapeutic cloning will advance scientific understanding of human development and disease, which may lead to treatments or cures for many people suffering from otherwise intractable or fatal disorders like those mentioned in chapter 2. However, before this technique can provide real clinical benefits for patients, many hurdles related to efficacy and safety must be overcome. There is much to learn about stem-cell biology and the mechanisms that regulate their self-renewal. No one knows if stem cells will differentiate properly after transplantation, form tumors, or otherwise develop inappropriately *in vivo.* Moreover, in order to maintain embryonic stem cells in culture in an undifferentiated state, they must be combined with mouse fibroblast cells. Until science develops another culture medium, the transfer of embryonic stem cells into humans is considered a xenograft and entails the risks of cross-species infection. Both of these uncertainties could entail harm for patients.

If these medical risks are overcome, the utilitarian can argue the appropriateness of human therapeutic cloning on two counts. First, since the early embryo has no capacity for pain, there is no utilitarian reason to prevent procedures that utilize it as a tool of science. If the overall outcome improves health without inflicting pain, utility deems the procedure ethical. Second, a utilitarian argument holds that any impropriety in the means is superseded by the significant benefits garnered from embryonic research. For the utilitarian, these arguments justify most procedures that alleviate suffering or increase overall pleasure.

In contrast, those who question the ethics of therapeutic cloning cite the comodification of the human embryo, the risk to egg donors, and justice issues as other facets of the technology that should be considered alongside potential outcomes. Therapeutic cloning involves the generation and destruction of the human embryo merely as a tool of science. The embryo is radically different from a sperm or egg or any other cell in the human body in that it possesses an independent internal code that directs its development into a human being. On this basis, it is objectionable to utilize it merely as a tool of research. Moreover, therapeutic cloning requires large numbers of eggs from multiple donors at the cost of $3,000 and upward in order to produce a viable clone from which to derive embryonic stem cells. Egg donation entails the stimulation of the ovaries with potent drugs followed by an invasive procedure to harvest the eggs. The long-term psychological and medical risks to the women involved are largely unknown. The potential for coercion of women who are in need of funds to pay for an expensive education or feed a family is obvious. This leads to the worry that those who are least well off financially may become the means for wealthy individuals to realize the potential benefits of this procedure. Given the expense and the inefficiency of therapeutic cloning, in conjunction with the limits of an already financially strained health care system, it is hard to imagine that it will become a widely accessible treatment. The promise of benefit is alluring, but a narrow utilitarian analysis overlooks several moral issues that merit a place in the discussion.

This leads to a second frequent critique of utilitarianism, which maintains that, in theory, utility cannot always renounce unjust social distributions. The principle of utility calls for the greatest pleasure for the greatest number. As such, utilitarianism allows the interests of the majority to supersede those of the minority. It requires that value be dispersed according to net combined happiness, rendering unjust distributions of limited resources theoretically possible. For example, if the greatest aggregate happiness for society comes from the enhancement of intelligence for those who can pay rather than from gene therapy for the uninsured, one interpretation of utility holds that the right action is enhancement of those who can afford the procedure. This is because society stands to benefit from the contributions of those with enhanced intelligence and loses no public resources, whereas gene therapy is costly and benefits relatively few. A strict calculation of the benefit:harm ratio could exclude the poorest sector from the potential benefits of genetic technology.

Third, according to theologian Scott Daniels, the foundation of utility has "logical difficulties that prevent it from satisfying common-sense limits to genetic interventions."[6] According to Daniels, the minimization of pain fails to offer suitable limits and distorts the significance of other important ethical values. He argues against the utilitarian contention that a state of affairs with no, or minimal, pain and suffering is valuable for its own sake. He maintains that other values, such as dignity and the common good, are at least as significant as pleasure. Moreover, Daniels contends that the focus on minimizing pain could actually thwart the ability to make ethical distinctions in the arena of human genetics. For example, utility reveals ethical inadequacies when dealing with questions concerning human embryonic research. Since the early embryo has no capacity for pain, there is no utilitarian reason to prevent procedures that treat the human embryo as an object of research. If the goal is healing or improving the human species, the utilitarian can argue the appropriateness of human embryonic research on two counts: 1) that pain is absent or minimal, and 2) that any impropriety of the means is superseded by the potential benefits garnered from embryonic research. These arguments render justifiable most reasonable research procedures attempting to alleviate pain or increase overall pleasure. On this basis, utilitarian reasoning could sanction extensive genetic control over human DNA.

It is important to note that the contemporary version of utility discussed here deviates from the classical utilitarianism presented by Socrates in the *Protagoras*. In contrast, Socrates presented a very sophisticated theory of consequentialism that rests upon acknowledged assessments of pleasure as good and pain as evil. He equates pleasure with good "if one's pleasure is honorable," yet he adds that there are "some pleasures which are not good, and some pains which are not evil."[7] For Socrates, the determination of pleasure versus pain rests on the evaluation of future consequences and their ramification upon a shared concept of good for the individual and society. As was mentioned in the introduction to this work, modern-day culture wars manifest themselves in indistinct convictions. Present disagreements about values can make the determination of pleasure and pain, and the resulting determination of moral boundaries, extremely difficult or impossible.

BEYOND UTILITY

A reinterpretation of utility by R. M. Hare offers some direction. Hare proposes a restructured view of utility that he labels a "credible form of utilitarianism."[8] His restructured version of classic utilitarianism holds that moral issues must be decided by rational tests. As such, he maintains that moral action is a special form of rational action. Two tenets provide the framework for his ethical theory. First, Hare relies on the study of moral language to reveal the formal properties of moral concepts. Second, he maintains that prescriptivity (favoring desires or likes) and universalizability (being unanimously valid) are central features of all moral judgments. From these precepts, he deems it possible to prescribe universally for all similar situations and assign equal weight to the interests of all affected participants.

How is this reasoning utilitarian? In order to assign equal weight to all parties in a situation, Hare considers the benefit or harm caused to one individual of equal value as the benefit or harm caused to another: the measure of utility includes the reduction of harm. He maintains that total benefit over the entire population is maximized if the interests of all parties are promoted, while weighing each equally. This is an interpretation of utility that requires maximizing not total utility but average utility. Hare revises the terms of benefit and harm to extend his theory of ethics beyond utility; his version adapts a formulation of the golden rule in lieu of the benefit:harm ratio. Hare's ethic requires that persons do for others what they would wish to be done for themselves were they to have the other's liking. In order to give equal weight to each individual's interest, including one's own, the ethical agent must place herself in all roles of the hypothetical situation. In this procedure, one's interests have the same weight as those of all other affected parties.

While the similarities to Rawls' "ideal observer" theory of justice are obvious, Hare differentiates himself from Rawls on several points.[9] While they both place the ethical agent in a hypothetical circumstance with some restrictions, Rawls speculates about what the ethical agent would do, while Hare's ethical agent takes action.

Otherwise, the hypothetical circumstances and the stated restrictions are formally analogous. Hare achieves the impartiality required by Rawls' "veil of ignorance," which requires lack of knowledge of social position in the formation of policy and the rationality of the ethical agent through formal means.

Another perceived disparity exists. Utility was cited above for its indifference to just and unjust distributions of benefit, as long as total utility is maximized. Hare argues against this allegation, citing that it ignores two important utilitarian considerations. First, attempts toward equality and justice increase total utility, thereby necessitating the inclusion of fairness in the calculus. Second, the converse also is relevant. Inequalities and injustices cause jealousy and resentment, culminating in overall disvalue. If the calculation includes these two factors, Hare maintains that the allegation that utility discounts inequalities of distribution is unfounded. For example, fantastic hypothetical cases, such as the cloning case discussed earlier, are unethical when evaluated via Hare's version of utility if the plight of the other and the resulting bitterness are assigned sufficient weight in the utilitarian ratio. In addition, if the ethical agent places himself or herself in all roles, the injustice of human cloning is readily apparent.

However, several problems remain, especially as one relates Hare to the specifics of inheritable genetic modification. Recall that Hare casts his ethic in terms of universal prescriptivism. The notion of universal prescriptivism narrows the overall flexibility of the principle of utility, by necessitating the calculation of the implications of universalizing inheritable genetic modification. The principle of utility generally applies only when consequences are predictable with some accuracy. Universal prescriptivism also requires the consideration of the collective desires or likes of all involved, actually and hypothetically. Extrapolating these requirements to future people's preferences for desirable genomes presents two problems for Hare. First, how are actual present desires balanced against hypothetical future desires in uncertain situations? Second, this balance, whether prescribed from one's own or from another's interests, must account for not only what others currently desire but also for what they should desire in unknown future situations.

In response, Hare maintains that the balance must be considered rationally and prudentially, with a clear assessment of what the consequences of the prescription would be if all were fully informed and unconfused. This entails universalizing prudence as the basis of the moral life. As such, the moral agent must make decisions considering the desires of others on equal terms with his or her own, and considering what those desires would be if all were perfectly prudent. This requirement is problematic in an ethical domain that includes only contemporaries. With future persons added to the equation, the task is virtually insurmountable.

The contributions and difficulties of the principle of utility as an ethical standard for inheritable genetic modification are obvious. Utility brings an emphasis upon respect for individual autonomy and the avoidance of aggregate harm to the other that is crucial in the assessment of inheritable genetic modification. Yet, utilitarianism lacks the capability to determine what constitutes a benefit or harm for future generations, and fails in its attempt to balance the actual interests of present persons against the hypothetical desires of future existents. As such, utility has difficulty determining where to go or when to stop. Thus, the question still remains: By what standards are we to assess inheritable genetic modification and its ramifications or identify when it is beyond our ability to evaluate? It is clear that there is not a single theory of ethics upon which to appraise genetic changes that impact future generations.

NOTES

1. John Stuart Mill, *Utilitarianism* (New York: MacMillan Publishing Company, 1957); and Jeremy Bentham, *An Introduction to the Principles of Morals and Legislation* (New York: Hofner, 1948).
2. Mill, *Utilitarianism*.
3. See Nicholas Capaldi, "A Catholic Perspective on Organ Sales," *Christian Bioethics* 2 (August 2000): 139–151.
4. American Society of Human Genetics Social Issues Subcommittee on Family Disclosure, "ASHG Statement, Professional Disclosure of Familial Genetic Information," *American Journal of Human Genetics,* no. 62 (1998): 474–483.
5. Daniel Callahan, "Manipulating Human Life: Is There No End to It?" In *Medicine Unbound,* ed. Robert H. Blank and Andrea L. Bonnicksen (New York: Columbia University Press, 1994), 118–131. Callahan further elucidates this point: "We can hardly know in advance with any certainty whether our manipulations will either achieve their intended aims; or if they do so, whether they will satisfy the desires or interests that led us to undertake them; or, moreover, whether they will do so without producing undesirable

and unforeseen side effects. It seems eminently reasonable, therefore, to make an advance judgment about the likelihood that a manipulation will work, both in the sense of achieving its intended goal and also leaving us satisfied that the goal was worth achieving" (121).

6. Scott Daniels, "Justifying Human Gene Therapy: An Assessment of Some of the Central Ethical Considerations Underlying the Application of Genetic Knowledge to Human Subjects from the Perspective of the Traditional Conscience" (Ph.D. diss., University of Tennessee, 1989).

7. Plato, "Protagoras," in *The Collected Dialogues of Plato*, ed. Edith Hamilton and Huntington Cairns (Princeton: Princeton University Press, 1961), 845–919.

8. R. M. Hare, "Ethical Theory and Utilitarianism," in *Utilitarianism and Beyond,* edited by Amartya Sen and Bernard Williams (New York: Cambridge University Press, 1982).

9. John Rawls, *A Theory of Justice* (Oxford: Oxford University Press, 1971).

AN ETHICAL MATRIX

The final section of this book proposes an ethical matrix to guide deliberations regarding inheritable genetic modification. The premise that spurred the formation of the matrix is that challenges posed by this technique are different and require an expansion of the ethical domain. The distinguishing aspects of purposeful inheritable genetic modification include intergenerational consequences and changes that involve an unconventional degree of specificity and control. Based on these characteristics, this technique raises fundamental questions about what it means to be human in the present and the future.

The task of including future persons in the ethical domain has only recently come into the scope of ethics. With the advent of a technology that impacts our progeny directly, it seems clear that future persons ought to have some claim on the moral agency of those who exist presently, but it is unclear what that claim entails. Therefore, the challenge is to formulate an expanded ethical framework to direct deliberation on a technique more powerful and far-reaching than anything achieved by science thus far. As science fiction becomes a scientific possibility, the interplay of cautiousness and daring witnessed in the public and professional sectors could cause divisions that obstruct the potential benefit of this new technique. The summons is to put the relationship between human dignity and biotechnology in perspective.

This goal is laden with several reservations. First, any attempt to assess the implications of far-reaching procedures and policies pertaining to inheritable genetic modification is fraught with empirical uncertainties. It is difficult, and probably inconceivable, to predict the ramifications of altering the human genome with any precision. Second, while it may be possible to ascertain that certain procedures are clearly ethical and others are clearly unethical, there will always be those procedures that are ethically questionable. Such determinations will remain exceedingly difficult. These perplexing cases are those in which a detailed analysis based upon pertinent ethical virtues, duties, and principles that ground human dignity could be especially beneficial. Although uncertainties cannot be eliminated, it is necessary to expand the present ethical domain to acknowledge the limit of human prognostication and address the implications of present actions upon future generations. This book proposes an alignment of comprehensive ethical theories to address that task.

There are several possible objections to such an alignment. The most probable objection is that it is impossible to align elements of divergent and mutually exclusive ethical theories with the expectation of facilitating dialogue. The argument is that alignment increases confusion rather than clarifies the issues. After all, difficulties arise from conflicts among the different ethical commitments or value considerations that divergent ethical theories stress; they are not solved by them. My response to this objection cannot resolve all the difficulties but does address some of the concerns.

First, and most important, the ethicists chosen to identify the parameters of the matrix espouse a mode of moral reasoning grounded in dignity. They understand dignity in a fundamentally practical way that serves to acknowledge the morally salient features of inheritable genetic modification and the way in which persons' moral commitments are important to reasoned reflection. A shared orientation grounded in dignity emphasizes the common strengths of each theory, so that diverse forms of reasoning can contribute to a common understanding and a means of approaching the issues involved.

Second, I concur with Alex London's explanation that "mutually exclusive methods of inquiry can be seen as compatible and complementary modes of moral inquiry with different rhetorical strengths."[1] London presents a conception of practical ethics derived from Aristotle's *Rhetoric* that represents the complex means by which agents develop and hold moral commitments. Aristotle's rhetorical model demonstrates that diverse approaches to practical reasoning can provide reasoned reflection that facilitates change in moral commitments. The strength is "to bring out those elements of truth in their particularity and marshal them in support of specific moral judgments in light of the facts of a particular case."[2] This approach has import for an inclusive approach to ethics.

London claims that the Aristotelian concept of practical ethics provides a means for understanding the value of diverse ethical methods as a means of deliberation that utilizes practical strengths independent of substantive differences. He contends that because of their different practical strengths, rather than in spite of them, different ethical foci are most helpful to practical moral deliberation when they are "nested together in subtle ways" rather than fragmented in "tradition disputes." Experience suggests that this is the common approach of practical ethicists in clinical settings and everyday ethical agents in their approach to life's dilemmas. London reports that "in the minds of many the very idea of constructing a moral theory is synonymous with bringing particular and more general moral judgments into a state of reflective equilibrium."[3] The strength in this approach is the juxtaposition of a narrative with the theoretical mooring of the ancient and classical ethical theories. As such, the differences among ethical theories become the most relevant criteria for their inclusion in moral deliberation, in order to provide practical ethicists with a more unified, accurate, and inclusive ground from which to consider complex issues in the biotechnological age.

Another likely objection to the use of diverse ethical theories as complementary modes of inquiry is that they frustrate rather than foster the consensus for which they strive. Consensus refers to agreement among a number of persons; agreement can be unanimous or majority rule, but most often consensus implies unanimity. There is no reason that diverse methodologies grounded in the concept of human dignity cannot reach consensus via different paths. Admittedly, it is equally likely that the different foci of diverse methods of ethical reasoning could lead to opposite conclusions. The goal of consensus is not the stated aim of the matrix. Moral justification typically originates from a sound foundation, such as the principle of dignity, not from consensus. To impose a requirement of consensus on a tool designed to facilitate dialogue is to misconstrue its intent and the task of bioethics.[4]

Terrence Ackerman, a bioethicist who is considered one of the "fathers" of bioethics, stated that the purpose of bioethics is to foster "informed choices of action for dealing with medical situations which raise moral issues."[5] He went on to explain that bioethics does not provide *the* moral answers, but assists in the analysis that must precede the determination of the best course of action. Identifying relevant values and questions pertinent to inheritable genetic modification facilitates this goal. The stated purpose of the matrix is to guide a comprehensive analysis of a biotechnological technique that renders each ethical theory, in itself, insufficient. The work of the matrix is prior to identification of policy options or the creation of an acceptable policy, both of which must eventually draw upon some form of consensus. Its task is to focus an interdisciplinary dialogue concerning inheritable genetic modification.

Dialogue, as it is used here, does not only connote conversation consisting of words and the exchange of ideas. It should be understood on an individual, local, national, and universal level. On the individual level, dialogue is an encounter with the other that brings about personal self-understanding and growth. On a local or national level, dialogue analyzes and critiques our contemporary way of life and ultimately signifies a means to live together in a pluralistic and rapidly changing world. "Dialogue encompasses all dimensions of our being human; it implies a global, existential dimension and involves the human subject in his or her entirety."[6] Dialogue is communication of a comprehensive nature.

NOTES

1. Alex John London, "Amenable to Reason: Aristotle's Rhetoric and the Moral Psychology of Practical Ethics," *Kennedy Institute of Ethics Journal* 30, no. 4 (December 2000): 287–307.
2. London, "Amenable."
3. London, "Amenable."
4. See Tristram Engelhardt, "Consensus Formation: The Creation of an Ideology," *Cambridge Quarterly of Healthcare Ethics* 11, no. 1 (Winter 2002): 7–16; and Mark Kuczewski, "Two Models of Ethical Consensus, or What Good Is a Bunch of Bioethicists?" *Cambridge Quarterly of Healthcare Ethics* 11, no. 1 (Winter 2002): 27–36.
5. Terrence F. Ackerman, "Experimentalism in Bioethics Research," *Journal of Medicine and Philosophy,* no. 2 (May 1983): 169–180.
6. Cardinal Walter Kasper, "The Nature and Purpose of Ecumenical Dialogue," *Harvard Divinity Bulletin* 30 (Winter 2001–2002): 19–23.

9

Inheritable Modifications Matrix

> It appears to me that in Ethics, difficulties and disagreements are due to a very simple cause: namely, to the attempt to answer questions, without first discovering precisely *what* question it is which you desire to answer.
>
> —George Edward Moore, *Principia Ethica*

Let me now propose the Inheritable Modifications Matrix, an ethical perspective to guide deliberation on research, procedures, and policies involving this technique. The matrix is intended to serve as a deliberative tool for genetics professionals, institutional review boards, or ethics committees to determine why and when certain inheritable genetic modifications are ethical and for what reasons others are unethical. Nonspecialists, individuals, and public policy commissions are also potential users of the matrix. The fundamental question of this practical approach to ethics is how to engage the moral commitments of persons who differ widely in their socioeconomic, cultural, political, religious, and racial backgrounds. The response is to provide a vehicle to facilitate a dialogue broad enough to include the different ways that individuals approach practical reasoning and form ethical commitments. The concept of matrix was selected for this purpose because it provides both the breadth and the flexibility to accommodate the complexity of this topic.

The term "matrix" has two meanings. Matrix designates womb, the place or enveloping element within which something originates, takes form, or develops.[1] The connotation of womb is particularly appropriate for an ethical perspective addressing the origin and future of human life. The second definition describes matrix as a set of numbers placed in rows and columns in a rectangular array. A quantitative matrix uses two axes to display two or more entries on the x-axis and an equal number of elements on the y-axis. In addition, a matrix can include two or three spatial dimensions. The ethical matrix proposed here is not for quantitative use; the intention is not to reduce ethics to calculations, but to facilitate reflective moral deliberation. The matrix constitutes the intersection of virtues, duties, and principles that characterize dignity and the scientific components related to clinical applications of inheritable genetic modification. I will describe the scientific axis first.

THE SCIENTIFIC AXIS

The scientific axis consists of three different stages of development and two purposes that pertain to inheritable genetic modification. Inclusion criteria are genetic changes that entail the introduction of DNA into gametes or early embryos for therapeutic or enhancement purposes. These procedures are summarized in the Inheritable Modifications Chart in table 9.1. The chart draws from a typology devised

Table 9.1. Inheritable Modifications Chart

Type 1: Gamete (egg and sperm) modification for therapeutic purposes
The replacement of malfunctioning genes with correctly functioning ones in human reproductive cells for the purpose of prevention, cure, or amelioration of genetic disease.

Type 2: Early embryo modification for therapeutic purposes
The replacement of malfunctioning genes with correctly functioning ones in the pre-implantation human embryo for the purpose of prevention, cure, or amelioration of genetic disease.

Type 3: *In utero* fetal alteration for therapeutic purposes
The replacement of malfunctioning genes with functioning ones *in utero* for the purpose of prevention, cure, or amelioration of genetic disorder. This procedure is aimed at somatic cells, but may inadvertently affect the germ line.

Type 4: Enhancement
The alteration of reproductive cells, early embryos, or fetuses for the purpose of improving specific human characteristics above normal.

Type 5: Degradation
The alteration of reproductive cells, early embryos, or fetuses for the purpose of diminishing specific human characteristics, such as intelligence.

by Wivel and Walters,[2] which was developed in 1993 and focused upon the distinctions between somatic cell and germ line modification and their intended purposes. At that time, Wivel and Walters considered the technical feasibility and the ethical acceptability of human germ line modification from a theoretical perspective. The question that they contemplated was: Is there merit in continuing the discussion about human germ line modification? They concluded that the dialogue should continue. The Inheritable Modifications Chart proposed in this book differs in that it is nearly a decade later and the scientific and ethical questions surrounding germ line modification are no longer theoretical but tangible. The science has progressed and the chart reflects those changes. All categories in the chart are similar in that the intended changes ultimately affect germ line cells and thus are inheritable. They differ in that the first three categories are various stages of development at which inheritable modification for therapeutic purposes occurs, and the fourth and fifth categories are purposes for which it is performed: enhancement or degradation.

Type 1, gamete modification for therapeutic purposes, considers gene transfer into eggs or sperm to prevent or cure disease. Gametes differ from all other cells of the human body, in that they are haploid cells containing one copy of each of the twenty-three human chromosomes. The function of gametes is to convey one copy of DNA from each parent to their offspring and, subsequently, to future generations.

Type 2, early embryo modification for therapeutic purposes, focuses on inheritable genetic modification in the pre-implantation embryo to prevent or cure disease. The insertion of a gene into an embryo is conveyed to both somatic cells and germ line cells during development, altering the embryo and its descendants.

Type 3, fetal modification for therapeutic purposes, includes all post-implantation genetic alteration of the fetus that is intended to prevent or cure disease. *In utero* fetal alteration usually occurs after the cells begin to differentiate into the various organs and systems that comprise the human body. Therefore, most changes attempted at this stage are intended to alter only the targeted cells. Type 3 qualifies for inheritable genetic modification since changes at this stage or later could inadvertently alter the germ line.[3] Type 3 is the only form of inheritable genetic modification that is, at the time of this writing, under consideration for research trials with human fetuses.[4]

The purpose or intended outcome of inheritable genetic modification distinguishes the two subsequent categories. Type 4, enhancement, may encompass the most difficult challenge that this technique presents. Enhancement is genetic alteration for the purpose of improving a given characteristic such as physical function, intelligence, appearance, or stature beyond what is considered normal for members of our species. Even though the enhancement versus therapy distinction is difficult in many instances, it is nevertheless useful and is widely applied in medical practice and health insurance contexts, as well as in our everyday judgments.

Type 5, degradation, is genetic alteration for the purpose of diminishing a distinctively human characteristic, such as intelligence, below what is considered normal for members of our species. The purpose is to produce humanlike creatures with lessened intelligence to perform undesirable and perilous tasks for society. Procedures that diminish human traits are clearly unethical in that they violate dignity and thwart participation in the human purpose.

Human Reproductive Cloning

Human reproductive cloning is widely discussed in popular literature and professional journals. It is the generation of a baby who is a time-delayed twin of its adult predecessor; genetically, the two are nearly exact replicas. The Inheritable Modification Chart excludes reproductive cloning since, technically, it is the replication of a genome, not the alteration of DNA. Human reproductive cloning is presently banned in the United States. At the time of this writing, it is considered medically irresponsible, based upon risks that include premature death and morbid obesity in animal models. However, as the science continues to improve, outcomes show increased success. The ethical concerns surrounding reproductive cloning cause a majority of Americans to oppose its use in humans. Those concerns include psychological problems for clones, plus far-reaching familial and societal issues. The multigenerational ethical issues raised by cloning are analogous to those engendered by inheritable genetic modification. Both techniques impact the genome of future persons and both involve questions of the objectification and commodification of human life. Therefore, the matrix could appropriately facilitate an ethical discussion on the topic of cloning, despite the technical discrepancy in procedural methods.

THE ETHICAL AXIS

The ethical axis of the Inheritable Modifications Matrix consists of a series of seven questions drawn from the virtues, duties, and principles identified by the analysis performed in part II. Questions are the tools of ethical inquiry, and the seven questions in the matrix assist moral deliberation by using the characteristics of dignity to guide discussion. There is precedent for this format in ethical analysis on scientific research. The Recombinant DNA Advisory Committee (RAC) of the National Institutes of Health (NIH) created the prototype of the ethical axis, in which it posed seven questions pertinent to the review of protocols for the transfer of DNA into the somatic cells of human subjects.[5] The questions posed by the matrix are similar to those written by the RAC, in that both documents raise issues of harm, benefit, justice, and informed consent. The questions proposed by the matrix differ in that they focus on purposeful inheritable genetic modification. The goal is to broaden the scope of the RAC questions to encompass the intergenerational aspects of this technique. The questions that comprise the ethical axis of the matrix follow. The explanation of the questions is brief, since part II covered the derivation of each.

Benefit versus Harm

The benefit versus harm ratio poses the question: *Does this research, procedure, or policy promote human health and well-being and minimize suffering?* Those procedures that are beneficial serve individuals and society by diminishing suffering and increasing well-being. Those procedures are harmful that result in overall suffering, detract from participation in the human purpose, or produce benefit at great personal or societal expense. The benefits ensuing from inheritable genetic modification must outweigh the potential harm involved in order for a procedure to be ethically justifiable.

Justice

The question of justice asks: *Will this procedure, research, or policy eventually provide equitable access and treatment to those suffering from genetic disorders?* The virtue of justice calls for the genetics professional to consider the needs of all persons who suffer from genetic disorders. A technology is not just if it neglects the poor or vulnerable,

Table 9.2. Inheritable Modifications Matrix

<table>
<tr><th></th><th></th><th>Gamete Alteration</th><th>Embryo Alteration</th><th>Fetal Alteration</th><th>Enhancement</th><th>Degradation</th></tr>
<tr><td rowspan="7">Human Dignity</td><td>Benefit</td><td colspan="5">Does this procedure/research/policy promote health and well-being and minimize suffering?</td></tr>
<tr><td>Justice</td><td colspan="5">Will this procedure/research/policy eventually provide access and treatment to those suffering from genetic disorders?</td></tr>
<tr><td>Responsibility</td><td colspan="5">Does this procedure/research/policy ensure the continuation of genuine human life (Jonas, The Imperative of Responsibility, 1984)?</td></tr>
<tr><td>Difference</td><td colspan="5">Does this procedure/research/policy seriously value circumstantial and genetic differences as essential to human flourishing and survival?</td></tr>
<tr><td>Integrity</td><td colspan="5">Does this procedure/research/policy demonstrate trustworthiness and honesty in inheritable genetic modification?</td></tr>
<tr><td>Discourse</td><td colspan="5">Is the discourse of the individual(s) involved a determining factor of the procedure/research/policy?</td></tr>
<tr><td>Wisdom</td><td colspan="5">Are the ultimate ramifications of this procedure/research/policy considered and evaluated?</td></tr>
</table>

or if it widens the gap between the haves and have-nots. The assessment of this question should encompass future considerations. Some procedures that are not accessible in the initial trial phases may satisfy the virtue of justice if they ultimately prove to be medically beneficial (in 70 percent or more of all cases) and are covered by health insurance as standard therapy.

Responsibility

The question of responsibility asks: *Does this procedure, research, or policy ensure the continuation of genuine human life?*[6] For Jonas, the imperative is to assure that human life, as we know it, continues. The second formulation of his Imperative of Responsibility further clarifies this duty: "Act so that the effects of your action are not destructive of the future possibility of such life."[7] Responsibility focuses primarily on the collective level. Inheritable genetic modifications that strive to ensure the continuation of human life elicit a positive response to this question. Procedures that threaten to destroy or drastically alter life evoke a negative response.

The Imperative of Responsibility has a greater scope than human genetics. In order for human life to continue, the breadth of the duty of responsibility must include the environment that sustains life.

> It is not mistaken moralism to underline that every person finds himself/herself in a context of obligations. None of us live for ourselves, with no relations to fellow-beings and to the environment around us. Morally, we are responsible for our actions and non-actions, as well as for the society in which we live.[8]

This book centers on inheritable genetic modification in humans because there is a tremendous burden of responsibility in this area. However, our duty as stewards extends far beyond anthropocentric concerns to the protection and preservation of the environment.

Difference

The question of difference asks: *Does this procedure, research, or policy seriously value circumstantial and genetic differences as essential to human flourishing and survival?* To espouse the ethic of the Other is to acknowledge the difference present in the face of the other person. This question addresses respect for diversity as essential for the advancement of the quality of life and the continuation of human existence. It applies to procedures that duplicate or eliminate a given genotype. Also, this question refers to planned "otherness." Beings who are truly "other" cannot be fashioned according to human design. It is possible to misconstrue the principle of difference to justify alterations that increase diversity beyond the point of human flourishing. If a procedure or research increases diversity (such as by generating quasi-humans) to the detriment of humankind, the answer to this question is "no." Moreover, the analysis of the other six questions in the matrix will prompt negative responses, clearly signaling the immoral nature of the act. If the procedure or research reduces diversity to the benefit of humankind, such as by eradicating a disease, the answer to this question is "yes," and the remaining six questions will affirm the morality of the act.

Integrity

This virtue asks: *Does this procedure, research, or policy demonstrate trustworthiness and honesty in inheritable genetic modification?* A person of integrity builds relationships of trust; she places the interest of the subject before personal or professional gain. A tremendous inequality in knowledge and power exists between the genetics professional and the subject of a procedure or research trial. Therefore, reliance on the trustworthiness of the professional is critical. This question refers especially to those inheritable genetic modifications that are lucrative or prestigious for the scientist or biotechnology company and potentially treat human life as a consumer commodity.

Discourse

The question of discourse asks: *Is the discourse of the individual(s) involved a determining factor of this procedure, research, or policy?* On the individual level, this question focuses on the process of dialogue that culminates in educated understanding as a prelude to decisions concerning inheritable genetic modification. In order to be consistent with Levinas's ethic, the voice of the proximate Other must be heard. One should not speculate about the saying of the Other, but must listen responsively to the Other in individual and collective situations.

On the collective level, it is necessary to recognize that there are variations in social or ethnic groups that may impact how they work through the ethics of a given question. When considering identifiable communities or subgroups, it is necessary to understand the different, informal decision-making processes that are operative in face-to-face dialogue. A process that is suitable for European-American populations may not be culturally appropriate for others. Clearly, communal discourse is more workable in smaller groups that are fairly homogeneous and becomes more problematic in larger, more detached populations.

The charge of proximate discourse becomes exponentially more challenging as the ability becomes available to make determinations for the Other of the distant future. The ethic of the Other, at a minimum, asserts the charge: "Do unto others as they would have you do unto them."[9] There are almost no circumstances in which we can know the wishes of another individual in the absence of proximate discourse.

Wisdom

The virtue of wisdom asks: *Are the ultimate ramifications of this procedure, research, or policy considered and evaluated?* Ultimate ramifications refer to the effects of actions on the collective level and in the future. What are the social and biological implications for persons and the planet? Wisdom is the virtue that orders all other virtues and discerns what is good or bad for humankind.

The second component of the question draws upon the virtue of humility. Humility in the context of inheritable genetic modification refers to the recognition of the magnitude of human dominion, rather than to the insignificance of human capabilities. It acknowledges the imbalance between what we can do scientifically and what we can evaluate ethically.

The Questions

Questions are often more helpful than answers. The seven questions presented in the matrix proceed from the characteristics of dignity pertinent to the endeavors of purposeful inheritable genetic modification. The order in which the questions are addressed has no theoretical implications. The matrix begins with benefit/harm and follows with justice since these two deliberations most efficaciously reveal those procedures that are ethically indefensible. Many of the quandaries concerning this technology will fall into the ethically dubious category. The remaining five questions are especially important in the evaluation of those procedures that involve subtle ethical distinctions. At each step, a negative response prompted by the matrix should indicate that the procedure, research, or policy under consideration is unethical based on the notion of dignity. The intervention should then be rejected, replaced, or reconsidered. The profundity and longevity of inheritable genetic modification warrants this cautious approach.

A CASE STUDY

Consider the following fictitious case to exemplify the use of the matrix. The case involves a first-trimester fetus diagnosed by DNA analysis with sickle cell disease (SCD). SCD is an autosomal recessive mutation (two copies of the mutated gene must be present to cause disease) that distorts red blood cells, making them relatively inflexible and rendering them unable to move through the small capillaries. The result is that the sickled cells plug up circulation and cause severe bone pain and, over the long term, lead to damage to internal organs, especially the heart, lungs, and kidneys. SCD patients are chronically anemic, which contributes to cardiac stress and strokes. The severity of symptoms fluctuates extensively in individual patients from relatively mild to very severe. The disease results in an average life span of forty years and decreased reproductive potential.

The incidence of SCD in persons of African descent is approximately 1 in 500 births. Carrier (one copy of the mutated gene and one functional copy) frequency is approximately 8 percent in the black population. Carriers of the sickle cell gene are moderately protected from the most severe and lethal consequences of malaria. This advantage apparently allowed carriers of this gene to be selected for over the long term in areas of the world where malaria is endemic, even though homozygotes (two copies of the mutated gene) are selected against. The following SCD case will serve as a model of a discussion about the ethical merits and pitfalls of inheritable genetic modification.

> A young couple, the Jacksons, learned that they both carry the gene for SCD after their son was born and diagnosed with the disorder. The Jacksons never considered the possibility of being carriers, despite the fact that Mr. Jackson's young-adult nephew is currently suffering from SCD. When Ms. Jackson became pregnant with their second child, the couple requested genetic testing and learned that the fetus also has SCD. Given their experience with their son's suffering and the imminent demise of their nephew, the couple wants to ensure that their second child is free from SCD. They are opposed to selective abortion and, therefore, are requesting genetic alteration of their fetus to correct the SCD gene.

The Discussion

For the purpose of this discussion, imagine that it is 2010 and a gene vector makes site-specific gene replacement surgery possible. The replacement of a gene at a specific site, as contrasted with crude vectors that add a gene at a random location in the genome, greatly increases the possibility of a good outcome for disorders like SCD that involve a single gene and nearly complete penetrance. Results in extensive animal trials and a research protocol involving humans substantiate this claim. Common complex disorders and traits that involve the interplay of many genes plus environmental factors are still difficult, if not impossible, to alter in a safe and efficacious manner. Despite the advances in gene surgery, ethical and medical questions about inheritable modifications remain.

The Jacksons' genetic counselor suggests that they use the matrix to guide their decision about genetic modification. The genetic alteration that the Jacksons are requesting meets the criteria for Type 3 on the scientific axis of the matrix (*in utero* fetal modification for therapeutic purposes). They begin the discussion by considering the first question of the ethical axis of the matrix, which is the benefit versus harm ratio of replacing the SCD gene in their fetus. The Jacksons know firsthand that the harm of nonintervention is serious and almost certain: SCD is extremely painful and ultimately fatal. However, the severity of the pain is unpredictable and the lifespan can vary from childhood to seventy years of age. Accordingly, the risks of genetic modification must be weighed against a life of sickness and premature death, and the outcomes of both options are uncertain. If the procedure is successful in replacing the gene, the potential benefit of a life free from the suffering of SCD is great. If it is not, the parents realize that this is a fetus that is already exposed to harm from the SCD gene. Based on a careful assessment of the risks and benefits, including available research data, all concur that the potential benefit to the fetus outweighs the possible harm of genetic modification, and they answer the question yes. On the other hand, if the risk:benefit assessment predicted greater harm than benefit, the treatment plan should be rejected or reformulated.

The question of the principle of *justice* is the next question posed by the matrix. This question raises issues of fairness. Equitable access to healthcare is a moral imperative and an important component of equal opportunity. In this 2010 scenario, inheritable genetic modification is still considered experimental. For this reason, it is not covered by insurance and is not generally accessible. Is it, therefore, unjust? The genetic counselor explains that there is a research trial approved and funded by the National Institutes of Health (NIH) that enrolls participants who meet criteria and provide informed consent. The intent of the research is to improve the safety and efficacy of genetic modification for SCD and eventually approve it for general use. Moreover, those who participate in the research contribute to the general knowledge that will potentially benefit themselves and others. Given this status, the procedure is just, and they answer the question yes. If access to the treatment were perpetually limited only to those who could pay, the procedure would be unjust because it would exclude those who are unable to afford its potential benefits.

The Jacksons move on to the question of *responsibility,* which refers to the duty to ensure the continuation of human life. Considering the individual level first, all concur that the amelioration of genetic disorder is a means of preserving human existence in that it promotes rather than thwarts the conditions for the possibility of future life. On the species level, the SCD gene adds a different dimension to the deliberation. From an evolutionary perspective, a single copy of the SCD gene confers an advantage on carriers that resulted in the high frequency of the gene in the black population over time. Recent evidence reports a significant reduction in mortality from all-cause mortality and severe malaria between the ages of two and sixteen months in malaria-endemic regions.[10] Is it then responsible to eradicate the gene on a large scale? The participants acknowledge that this advantage is an important factor in reducing morbidity in certain regions but agree that it is not a significant factor in the United States, where the incidence of malaria is very low and treatment is available. Consequently, they answer the question of responsibility yes.

The matrix question of *difference* is germane to this case, since the goal of the procedure is the elimination of the SCD gene from the Jacksons' descendants and, potentially, from a large segment of the population if this technique were ever widely implemented. In this case, since the difference under consideration is a disease gene that thwarts human flourishing and survival, the participants answer the question yes. If the effect were to eliminate a difference that was essential to human flourishing and survival, the answer to the question would be no.

The question of the virtue of *integrity* follows. In response to this query, the deliberations conclude that the intent to cure a genetic disorder is an honorable purpose. The amelioration of fatal disease characterizes the traditional goals of medicine and attests to the trustworthiness of the professionals and others involved. The question is answered yes.

The discussion next turns to the question of *discourse*. It is obvious that the consent of future persons is not possible in all procedures involving inheritable genetic modification. It is customary for parents or guardians to make medical decisions in the best interest of children or those who are incompetent, but there are almost no instances in which best-interest judgments can be made for future generations. However, the case of a fatal disorder such as SCD may be an exception. It is virtually certain that the eradication of fatal genetic disease serves the best interests of future persons. There is precedent for a decision to treat patients in the absence of informed consent in standard medical practice. In life-threatening emergency situations, treatment is begun without consent if the patient is unable to give consent and the family is unavailable, or if time is of the essence. It is not only ethically acceptable, but ethically obligatory, to waive informed consent in life-threatening emergencies. In the case of inheritable genetic modification, the effects of the procedure are different for two reasons: the situation is not usually emergent and the alteration is not limited to the person being treated. As such, inheritable genetic modification "raises the bar" for therapy without the consent of the patient. In order to alter the genome of a future person, there must be virtual certainty that she would approve the procedure. For a fatal disorder like SCD, which involves significant suffering at best and extreme suffering at worst, that assurance exists. Thus, the question of discourse elicits an affirmative response.

The final question of *wisdom* raises an issue that underscores the virtue of humility and the limits of human prognostication. As mentioned, SCD displays extreme variability in the range of symptoms and the longevity experienced by patients. Until scientists identify genetic markers that predict severity, there is no way to know whether the future child will suffer from severe pain and early death or experience controllable pain and live to seventy years of age. In light of this uncertainty, it is extremely difficult to calculate the wisdom of tampering with the genome to replace the SCD gene. In this case, the question of wisdom accentuates the discrepancy between what we can do scientifically and what we can evaluate ethically. As a result, the Jacksons determine that they cannot answer the question with a simple yes or no. They decide to rely on the other six questions to determine the ethics of the procedure. Based on the positive responses to six of the questions, and in the absence of a definitive no to the question of wisdom, they concur that the correction of the SCD gene is overall an ethically and medically responsible procedure.

Next, imagine that the year is 2020 and the standard for acceptable babies is higher. Diagnostic "microarray chips" can test DNA for thousands of genes or all the genes simultaneously and predict predisposition to common diseases and traits based on the results. Now parents not only want babies that are free from lethal disorders, but they want their babies free from even a significantly higher than average risk of disease or disability. Let us fast-forward the Jacksons' case to 2020.

Since the Jacksons' fetus will be modified for the SCD gene, the couple decides that they want to alter their fetus to enhance the possibility of a good life. Mr. Jackson suffered from learning disabilities as a child, and his firstborn already exhibits these same traits. Due to the burden of his

> disability and the resulting loss of opportunity, the father wants to correct the malfunctioning gene for learning in the fetus. He does not want the child's intelligence enhanced above average, only the child's learning ability corrected to a normal level.

Since learning disabilities involve multiple genes and gene/environmental interactions, and since they are not fatal, they require further consideration. The genetic counselor suggests the appraisal of this new request through the matrix.

The discussion returns to the question of *benefit versus harm* of the procedure. This time the issues are different. The Jacksons want to alter a trait that is functioning, even if poorly. Without intervention, the child will experience difficulties in school, but otherwise will have a normal life; most individuals with learning disabilities do well, and many thrive. Conversely, the uncertainties and risks of genetically altering learning may be great. Many genes operate in conjunction with the environment to control learning. Thus, the medical outcome of such a procedure (even assuming safety) will always be uncertain. It is impossible to modify all of the genes involved with learning or to identify and control all of the aspects of the environment that impact learning. Moreover, there are many difficulties associated with complex gene regulation, and unexpected or undesirable results are likely. If the procedure were medically possible, an improvement in learning ability would clearly be a benefit to the child and possibly to its future progeny. However, given the risks, the benefit versus harm assessment concludes that the outcome of the procedure is uncertain at best, could cause greater harm than benefit, and therefore should not be performed. Nevertheless, the Jacksons persist with their request.

The deliberation proceeds to the question of *justice*. The correction of learning disabilities will not be covered by insurance and is inaccessible to most individuals.[11] On this premise, the question of justice is answered "no." However, this family has the ability to pay for the procedure. They contend that many parents presently enhance their children's ability to learn through private schools and tutors. They see no difference between tutoring to compensate for learning disabilities and correcting them through gene surgery. Disarmed by their cavalier response, the counselor encourages the Jacksons to continue the analysis through the other five questions of the matrix as a means of further assessing the ethics of their request to alter the learning disability genes.

The matrix question of *responsibility* follows. The modification of certain traits, such as intelligence, impacts that which is fundamental to being human. Enhancement of intelligence can be very beneficial on the individual level, but the issues on the societal level are different. Overemphasis of one or several traits can lead to forms of manipulation and control that alter human life in significant ways and impact individual freedom. Because of this serious concern, the group concurs that this procedure has potential ramifications that are beyond their ability to evaluate and judge responsibly, and they answer the question "no."

The question of *difference* ensues. The purpose of the procedure is to eliminate a learning difference that is a burden in the present but could potentially be beneficial in the world of the future. Moreover, it exerts pressure on others to do the same in order to "level the playing field." Those who are different and who cannot or will not submit to cognitive enhancement are left behind, while those who can afford it seek bigger and better genetic boosts. Furthermore, what if the altered child does not measure up to parental expectations? For example, what if the learning disabilities are corrected but the child exhibits violent behavior patterns or some other unexpected result of genetic alteration? Will the parents accept such a child, or are they demanding only perfection? Given these concerns, the response to the question of difference is another "no."

The question of the virtue of *integrity* generates a similar conclusion. Cognitive alteration could objectify human life or reduce it to a vehicle of profit. Moreover, a physician with integrity should not agree to perform this alteration based on uncertain medical outcomes and ethical problems, despite professional or monetary gain.

The consideration of *discourse* focuses attention on the reality that the consent of the child obviously cannot be obtained for a procedure that will fundamentally change it and all future descendants. This question prompts a negative answer since it is not an emergent or life-threatening disorder for which consent is waived, and it is impossible to conceive what style of learning a child will desire in future societies.

The final question of *wisdom* confirms that this procedure is inadvisable, since its ultimate implications are clearly uncertain and potentially ominous. This analysis reveals that it is impossible to evaluate with any level of certainty the effects of altering learning ability for the individual or society. Given that six of the seven questions triggered negative answers, the genetic counselor states that it is medically and ethically inappropriate to perform an inheritable genetic procedure to avoid learning disabilities. The couple remains unconvinced and requests that the ethics committee utilize the matrix to reconsider their request.

It is now 2030 and the trend in assisted reproduction is toward even higher standards. Parents now want babies free from "undesirable" traits or characteristics. They are demanding and willing to pay for perfect heath, superior intellect, beauty, physical strength and stature, a good disposition, and more. The market for these modifications is booming, but they are not covered by a severely stretched health insurance system. As a result, widespread use of genetic technology has gradually increased the average level of health and well-being for the wealthy, while the poor are left behind. The medical and ethical conundrums are more profound than ever.

The inheritable genetic modifications discussed here cover only a fraction of the ethical spectrum. Our genetic options will increase dramatically with each new scientific advance. The reasons for endorsing or rejecting procedures at either extreme of the spectrum are clear. Alterations that fall into the ethically dubious zone require an expanded discussion that involves the public and representatives of diverse disciplines. The intent of the matrix is to provide a mechanism to focus the moral dialogue of decision-making groups and individuals faced with the difficult challenge of assessing procedures, research, and policies involving purposeful inheritable genetic modification.

EPILOGUE

Advances in molecular genetics exponentially increase the capacity to transform our world and modify that which is uniquely human. The power residing in the new genetics tends to blur the distinction between that which is a magnificent gift from God and that which is a product of technology. Scientifically, we can accomplish increasingly greater and greater feats. However, the reality that we "may" be able to accomplish something should not elicit a "must do" mandate. The delineation of limits for this new technique is material for reflection.

There are numerous examples in everyday and clinical experience of exciting new scientific discoveries that begin with a limited purpose and quickly extend into general use and occasionally overuse. For example, the drug Viagra was developed to treat heart conditions and only inadvertently turned out to be an effective treatment for male sexual impotency. It almost automatically expanded from a therapy to widespread use for enhanced male sexual performance. The success of Viagra in males subsequently prompted the production of similar enhancements for females. A therapeutic drug quickly evolved into an enhancement and, subsequently, has raised the standard for acceptable sexual performance in the general population. Based on the potential for biotechnology to follow a similar course, one can imagine a scenario in which no extent of genetic boost is ever sufficient. Heller summarizes the challenge:

> On what moral basis will we choose between the various options germline therapy makes available to us? Which, if any, genetic problems will we try to fix permanently for future persons; that is which problems will we try to eliminate from the gene pool and thus which persons will live and which will not? How will we decide such questions now that we are confronted with them, as perhaps we have never been before in the history of humankind?[12]

Can we safeguard what is uniquely human, recognizing that the alteration of the human genome is inevitable? Can we draw the line and sustain it?

The reflections on purposeful inheritable genetic modification expressed in this chapter point to an ethical boundary. For clarity, this project utilizes the clinical definition of genetic disorder as those malfunctions that

> include single gene or monogenic disorders, chromosomal disorders, congenital malformations, or other common diseases such as cancer in which genetic predisposition plays a large part (multi-factorial diseases), disorders of the mitochondria, and disorders due to random somatic cell mutations.[13]

Virtually every known disorder fits this classification, or will as more and more genes are linked to disease.

Since it is clear that not every genetic disorder is appropriate for this technique, a further qualifier is necessary. Due to the gravity and permanence of this technique, I propose the criterion of "fatal" as a further delineation for genetic disorders appropriate for inheritable modification once science resolves the issues of safety and efficacy in the procedure. "Fatal" refers to disorders leading to premature death, and "premature" indicates death that is untimely or prior to an average life expectancy. This can be established with a fair degree of clarity. Cystic fibrosis clearly brings about untimely death; Huntington's disease does also, despite its late onset.

The designation of fatal entails a causal connection of death with a given genotype. Those genotypes with low penetrance, or those disorders that involve multiple genes or environmental factors, do not presently qualify for inheritable genetic modification for medical reasons, even though they are fatal. Schizophrenia is an example of a catastrophic and potentially fatal disorder, but its pattern of inheritance and its environmental components make it an unlikely candidate for inheritable genetic modification at present. Early evidence suggests that several genes acting in concert with environmental triggers cause schizophrenia. The risk of developing schizophrenia for persons who likely have the gene(s) is approximately 12 percent, rendering the development of symptoms uncertain. The uncertainties inherent to schizophrenia, and to other genetic disorders with similar inheritance patterns, make it presently unsuitable for inheritable genetic modification. If scientific advances identify markers that predict onset and severity, it would become a likely candidate.

The stipulation of fatal genetic disorder is an extremely cautious approach to take to the potential benefits of inheritable genetic modification. The designation of fatal is more restrictive than limits suggestions by others, such as earlier stipulations of "severely debilitating disease" or even "fatal or devastating disease." My eliminating any reference to debilitating or devastating disorders is intentional: the term "fatal" precludes those disorders that are severely burdensome, but not life-threatening. While this may seem heartless or less than benevolent, it is due to the scientific and philosophical uncertainties that accompany the designation of "severe," "debilitating," and "devastating." What constitutes severity? When is a disorder debilitating? Who establishes the criteria for devastating? Does debilitating include emotional, spiritual, or physical disorders? These are ambiguities that are unsuitable in the context of an inheritable change that is, for all practical purposes, permanent.

The qualifier of fatal avoids many of the ambiguities alluded to above and establishes a boundary that can be verified in most instances. The possibility of a discernible boundary for this technique ceases to exist if the qualifier of fatal is diluted. The paradox of inheritable genetic modification is that it involves outcomes that are both enduring and unpredictable. This reality warrants an extremely exacting approach. Moreover, if the ethical questions raised in the matrix are even remotely appropriate, it is difficult to justify an expansion of the boundary of inheritable genetic modification beyond those genetic disorders that are life-threatening. Beyond that limit, human values that are important components of human dignity are endangered. Pioneer of gene therapy W. French Anderson acknowledged an "uneasiness" about inheritable genetic alteration and asserted that there is justification for drawing the line, whether it is considered from a "theological perspective or a secular humanist one."[14]

The designation of fatal genetic disorder is a beginning point. The final determination of the ethical boundary of inheritable genetic modification should include a process involving interdisciplinary moral dialogue and evaluation. The ramifications of genetic technology will be one of the defining characteristics of the twenty-first century. Ethical deliberation at all levels is a crucial part of the development and utilization of this powerful technique. This work has been presented in an effort to facilitate that process and assist in the identification of the boundary beyond which we must exercise the ethical "no."

NOTES

1. Noah Webster, *The Living Webster Encyclopedic Dictionary of the English Language* (Chicago: The English Language Institute of America, 1975).
2. Nelson Wivel and Leroy Walters, "Germ Line Gene Modification and Disease Prevention: Some Medical and Ethical Perspectives," *Science* 262 (22 October 1993): 533–537.
3. Eliot Marshall, "Gene Therapy: Panel Reviews Risks of Germ Line Changes," *Science* 294 (14 December 2001): 2268–2269.
4. See C. Coutelle and C. Rodeck, "On the Scientific and Emotional Issues of Fetal Somatic Gene Therapy," *Gene Therapy* 9 (June 2002): 670–673; and Esmail Zanjani and W. French Anderson, "Prospects for in Utero Human Gene Therapy," *Science* 285 (24 September 1999): 2084–2088.
5. Recombinant DNA Advisory Committee, National Institutes of Health, "Regulatory Issues, Selected Minutes from October 6, 1989 RAC Meeting," *Human Gene Therapy*, no. 1 (1990): 185–189.
6. Hans Jonas, *The Imperative of Responsibility: In Search of an Ethics for the Technological Age* (Chicago: University of Chicago Press, 1984).
7. Jonas, *The Imperative.*
8. T. Austad, "The Right Not to Know—Worthy of Preservation Any Longer? An Ethical Perspective," *Clinical Genetics,* no. 50 (1996): 85–88.
9. Jere Surber, *Ethical Principles: Levinas and the Tradition,* (Denver: University of Denver, 1996).

10. Michael Aidoo, Diane Terlouw, Margarette Kolczak, Peter McElroy, Feiko ter Kuile, Simon Kariuki, Bernard Nahlen, Altaf Lal, and Venkatachalam Udhayakumar, "Protective Effects of the Sickle Cell Gene Against Malaria Morbidity and Mortality," *The Lancet* 359 (13 April 2002): 1311–1312; and T. R. Kotila and W. A. Shokunbi, "Survival Advantage in Female Patients with Sickle Cell Anaemia," *East Africa Medical Journal* 78 (July 2001): 373–375.

11. It is probable that genetic alteration for enhancement purposes will perpetually be excluded from healthcare coverage. The expense of an enhancement procedure cannot be warranted by the cost of medical care over multiple generations, as is the case of inheritable genetic disorders. For this reason, access to enhancement will likely be limited to those who can afford to pay.

12. Jan Heller, *Human Genome Research and the Challenge of Contingent Future Persons* (Omaha, Nebraska: Creighton University Press, 1996).

13. Heller, *Human Genome.*

14. W. French Anderson, "Genetics and Human Malleability." *Hastings Center Report*, no. 20 (January–February 1990): 21–24.

Bibliography

Ackerman, Terrence F. "Experimentalism in Bioethics Research." *Journal of Medicine and Philosophy* 2 (May 1983): 169–80.

Almeida-Porada, C., et al. "Adult Stem Cell Plasticity and Methods of Detection." *Reviews in Clinical & Experimental Hematology* 5 (2001): 26–41.

Anderson, W. French. "End-of-the-Year Potpourri—1996." *Human Gene Therapy* 7 (1996): 2201–02.

———. "Genetics and Human Malleability." *Hastings Center Report* 20 (January–February 1990): 21–24.

American Society of Human Genetics Social Issues Subcommittee on Family Disclosure. "ASHG Statement, Professional Disclosure of Familial Genetic Information." *American Journal of Human Genetics* 62 (1998): 474–83.

Aristotle. *Nichomachean Ethics.* Translated by H. Rackham. Cambridge, Mass.: Harvard University Press, 1994.

Aquinas, Saint Thomas. *Summa Theologica.* Translated by the Fathers of the English Dominican Province. London: Burns Oates & Washbourne Ltd., 1953.

Asch, Adrienne. "Distracted by Disability." *Cambridge Quarterly of Healthcare Ethics* 7 (Winter 1998): 77–87.

Augustine, Saint. *The City of God.* Translated by Marcus Dods, New York: The Modern Library, 1993.

Austad, T. "The Right Not to Know—Worthy of Preservation Any Longer? An Ethical Perspective." *Clinical Genetics* 50 (1996): 85–88.

Barritt., J. A., C. A. Brenner, H. E. Malter, and J. Cohen. "Mitochondria in Human Offspring Derived from Ooplasmic Transplantation." *Human Reproduction* 16 (2001): 513–16.

Bentham, Jeremy. *An Introduction to the Principles of Morals and Legislation.* New York: Hofner, 1948.

Bostanci, Adam. "Blood Test Flags Agent in Death of Penn Subject." *Science* 295 (25 January 2002): 604–5.

Boyce, Nell. "Trial Halted after Gene Shows Up in Semen," *Nature* 414 (13 December 2001): 677–78.

Bryne, Peter. *The Philosophical and Theological Foundations of Ethics*. New York: St. Martin's Press, Inc., 1992.

Cahill, L. S. "Genetics, Commodification, and Social Justice in the Globalization Era." *Kennedy Institute of Ethics Journal* 11 (September 2001): 221–38.

Callahan, Daniel. "Universalism and Particularism." *Hastings Center Report* 30 (January–February 2000): 37–44.

———. "Manipulating Human Life: Is There No End to It?" In *Medicine Unbound*, edited by Robert H. Blank and Andrea L. Bonnicksen, 118–31. New York: Columbia University Press, 1994.

Capaldi, Nicholas. "A Catholic Perspective on Organ Sales." *Christian Bioethics* 2 (August 2000): 139–51.

Caplan, Arthur L. "If Gene Therapy is the Cure, What is the Disease?" In *Gene Mapping, Using Law and Ethics as Guides*, edited by George Annas and Sherman Elias, 128–41. New York: Oxford University Press, 1992.

Cavazzana-Calvo, Marina, et al. "Gene Therapy of Human Severe Combined Immunodeficiency (SCID)-X1 Disease." *Science* 288 (2000): 669–72.

Chambers, Tod. "Centering Bioethics." *Hastings Center Report 30* 1 (January–February 2000): 22–29.

Crosby, John F. "The Personhood of the Human Embryo." *Journal of Medicine and Philosophy* 18 (August 1993): 399–417.

———. "Person, Consciousness." *Christian Bioethics* 1 (April 2000): 37–48.

Daniels, Scott Eugene. "Justifying Human Gene Therapy: An Assessment of Some of the Central Ethical Considerations Underlying the Application of Genetic Knowledge to Human Subjects from the Perspective of the Traditional Conscience." Ph.D. dissertation, University of Tennessee, 1989.

Donovan, P. J. and J. Gearhart. "The End of the Beginning for Pluripotent Stem Cells." *Nature* 414 (1 November 2001): 92–97.

Dworkin, Ronald. *Life's Dominion*. New York: Vintage Books, 1993.

Engelhardt, Tristram. "Consensus Formation: The Creation of an Ideology." *Cambridge Quarterly of Healthcare Ethics* 11 (Winter 2002): 7–16.

———. *The Foundations of Bioethics*. New York: Oxford University Press, 1986.

Engelking, Constance. "The Human Genome Exposed: A Glimpse of Promise, Predicament, and Impact on Practice." *Oncology Nursing Forum* 22 (Supplement, 1995a): 3–9.

Fletcher, John C. "Evolution of Ethical Debate About Human Gene Therapy." *Human Gene Therapy* 1 (1990): 55–68.

Float, T. and B. Laube. "Gene Therapy in Cystic Fibrosis," *Chest* 120 (September 2001): 124S–142S.

Frankel, Mark S. and Audrey R. Chapman. "Human Inheritable Genetic Modifications: Assessing Scientific, Ethical, Religious, and Policy Issues." American Association for the Advancement of Science, September 2000, www.aaas.org/spp/dspp/sfrl/germline/main.htm.

Friedman, Theodore. "Principles for Human Gene Therapy Studies," *Science* 287 (24 March 2000): 2163–68.

Gustafson, James M. "Theology Confronts Technology and the Life Sciences." In *On Moral Medicine*, edited by Stephen E. Lammers and Allen Verhey, 35–40. Grand Rapids, Mich.: William B. Eerdmans Publishing Company, 1987.

———. "Christian Humanism and the Human Mind." In *On Moral Medicine*, edited by Stephen E. Lammers and Allen Verhey, 573–82. Grand Rapids, Mich.: William B. Eerdmans Publishing Company, 1987.

Hall, Mark A. and Stephen S. Rich. "Patients Fear of Genetic Discrimination by Health Insurers: The Impact of Legal Protections." *Genetics in Medicine* 4 (July–August 2000): 214–21.

Hare, P. M. "Ethical Theory and Utilitarianism." In *Utilitarianism and Beyond*, edited by Amartya Sen and Bernard Williams. New York: Cambridge University Press, 1982.

Hawthorne, Nathaniel. "The Birthmark." In *The Portable Hawthorne*, edited by Malcolm Cowley, 164–85. New York: Viking Press, 1948.

Heller, Jan Christian. *Human Genome Research and the Challenge of Contingent Future Persons*. Omaha, Nebr.: Creighton University Press, 1996.

Herman, Barbara. *The Practice of Moral Judgment*. Cambridge, Mass.: Harvard University Press, 1993.

High, K. A. "AAV-Mediated Gene Transfer for Hemophilia," *Annals of the New York Academy of Science* 953 (December 2001): 64–74.

Holland, S. "Contested Commodities at Both Ends of Life: Buying and Selling Gametes, Embryos, and Body Tissues." *Kennedy Institute of Ethics Journal* 11 (September 2001): 263–84.

Hollenbach, David. *Justice, Peace, and Human Rights*. New York: Crossroads, 1988.

Hull, David L. *Science as a Process*. Chicago: University of Chicago Press, 1995.

Jonas, Hans. *Philosophical Essays*. Englewood Cliffs, N.J.: Prentice Hall, Inc., 1974.

———. *The Imperative of Responsibility: In Search of an Ethics for the Technological Age*. Chicago: The University of Chicago Press, 1984.

———. "Ethics and Biogenetic Art." *Social Research* 52 (Autumn 1985): 491–504.

John Paul II. "The Ethics of Genetic Manipulation: John Paul II to Medical Association." *Origins* 13 (17 November 1983): 385–89.

———. "The Human Person—Beginning and End of Scientific Research," Address of Pope John Paul II to the Pontifical Academy of Sciences (28 October 1994). *The Pope Speaks* 40 (March–April 1995): 80–84.

Kaebnick, Gregory E. "On the Sanctity of Nature." *Hastings Center Report* 30 (September–October 2000): 16–23.

Kaji, E. H. and J. M. Leiden. "Gene and Stem Cell Therapies." *Journal of the American Medical Association* 285 (2001): 545–50.

Kant, Immanuel. *Groundwork of the Metaphysic of Morals*. Translated and analyzed by H. J. Paton. New York: Harper & Row, 1956.

———. *The Critique of Pure Reason*. Translated by Norman Kemp Smith. New York: St Martin's Press, 1965.

Kasper, Walter. "The Nature and Purpose of Ecumenical Dialogue." *Harvard Divinity Bulletin* 30 (Winter 2001–2002): 19–23.

Kass, L. "The Moral Meaning of Genetic Technology." *Commentary* 108 (1999): 32–41.

Keenan, James F. "What is Morally New in Genetic Manipulation?" *Human Gene Therapy* 1 (1990): 289–98.

Krause, D. S. et al. "Multi-Organ, Multi-Lineage Engraftment by a Single Bone Marrow-Derived Stem Cell." *Cell* 105 (May 2001): 369–77.

Kuczewski, Mark. "Two Models of Ethical Consensus, or What Good is a Bunch of Bioethicists?" *Cambridge Quarterly of Healthcare Ethics* 11 (Winter 2002): 27–36.

Levinas, Emmanuel. "Ethics as First Philosophy." In *The Levinas Reader*, edited by Sean Hand. Cambridge, Mass.: Blackwell, 1993.

———. *Collected Philosophical Papers*. Translated by Alphonso Lingis. Boston: Martinus Nijhoff Publishers, 1987.

———. *Ethics and Infinity*. Translated by Richard A. Cohen. Pittsburgh: Duquesne University Press, 1985.

———. *Otherwise Than Being or Beyond Essence*. Boston: Martinus Nijhoff Publishers, 1981.

———. *Totality and Infinity*. Translated by Alphonso Lingis. Pittsburgh: Duquesne University Press, 1969.

London, Alex John. "Amenable to Reason: Aristotle's Rhetoric and the Moral Psychology of Practical Ethics." *Kennedy Institute of Ethics Journal* 30 (December 2000): 287–307.

Lozier, Jay, et al. "Toxicity of First Generation Adenoviral Vector in Rhesus Macaques." *Human Gene Therapy* 13 (1 January 2002): 113–24.

MacIntrye, Alasdair. *After Virtue*. Notre Dame, Ind.: University of Notre Dame Press, 1984.

Mauron, Alex, and Jean-Marie Thevoz. "Germ-line Engineering: A Few European Voices." *The Journal of Medicine and Philosophy* 16 (December 1991): 649–66.

Marshall, Eliot. "Gene Therapy: Panel Reviews Risks of Germ Line Changes," *Science* 294 (14 December 2001): 2268–69.

McCreath, J., K.H.S. Howcroft, A. Campbell, A. Coleman, A. E. Schnieke, and A. J. Kind. "Production of Gene-Targeted Sheep by Nuclear Transfer from Cultured Somatic Cells." *Nature* 405 (29 June 2000): 1066–69.

Mill, John Stuart. *Utilitarianism.* New York: MacMillan Publishing Company, 1957.

Moraczewski, Albert S. "The Church and the Restructuring of Humans." In *Medicine Unbound,* edited by Robert H. Blank and Andrea R. Bonnicksen, 40–60. New York: Columbia University Press, 1994.

Morris, David. "How to Speak Postmodern: Medicine, Illness, and Cultural Change." *Hastings Center Report 30* 6 (November–December 2000): 7–16.

Murray Thomas H. "The Human Genome Project: Ethical and Social Implications." *Bulletin of the Medical Library Association* 83 (January 1995): 14–21.

Nelson, J. Robert. *On the New Frontiers of Genetics and Religion.* Grand Rapids, Mich.: William B. Eerdmans Publishing Company, 1994.

Niebuhr, Reinhold. *The Nature and Destiny of Man.* Volume 2, *Human Destiny.* New York: Charles Scribner's Sons, 1942.

Orlic et al., "Bone Marrow Cells Regenerate Infracted Myocardium." *Nature* 410 (2001): 701–5.

Pellegrino, Edmund. *The Christian Virtues in Medical Practice.* Washington, D.C.: Georgetown University Press, 1996.

———. "Toward a Virtue-Based Normative Ethics for the Health Professions." *Kennedy Institute of Ethics Journal* 5 (September 1995): 253–77.

———. "The Medical Profession as a Moral Community." *Bulletin of the New York Academy of Medicine* 66 (May–June 1990a): 221–32.

Peters, Ted. "In Search of the Perfect Child: Genetic Testing and Selective Abortion." *Christian Century* 113 (30 October 1996): 1034–37.

Plato. "Theatetus." In *The Collected Dialogues of Plato,* edited by Edith Hamilton and Huntington Cairns, 845–919. Princeton: Princeton University Press, 1961.

———. "Meno." In *The Collected Dialogues of Plato,* edited by Edith Hamilton and Huntington Cairns, 353–84. Princeton: Princeton University Press, 1961.

———. "Republic." In *The Collected Dialogues of Plato,* edited by Edith Hamilton and Huntington Cairns, 575–844. Princeton: Princeton University Press, 1961.

Poland, Susan Carter. "Genes, Patents, and Bioethics: Will History Repeat Itself?" *Kennedy Institute of Ethics Journal* 10 (September 2000): 265–81.

Rahner, Karl, S. J. *Christian at the Crossroads.* Translated by V. Green. New York: The Seabury Press, 1975.

———. *Foundations of Christian Faith.* Translated by W. D. Dych. New York: The Seabury Press, 1978.

Rawls, John. *A Theory of Justice.* Oxford: Oxford University Press, 1971.

———. "Social Utility and Primary Goods." In *Utilitarianism and Beyond,* edited by Amartya Sen and Bernard Williams, 159–85. New York: Cambridge University Press, 1982.

Recombinant DNA Advisory Committee, National Institutes of Health. "Regulatory Issues, Selected Minutes from October 6, 1989 RAC Meeting." *Human Gene Therapy* 1 (1990): 185–89.

Reya, T., et al. "Stem Cells, Cancer, and Cancer Stem Cells." *Nature* 414 (2001): 105–11.

Ross, Gail, et al. "Gene Therapy in the United States: A Five-Year Status Report." *Human Gene Therapy* 7 (1996): 1781–90.

Shapiro, Harold T. "Reflection of the Interface of Bioethics, Public Policy, and Science." *Kennedy Institute of Ethics* 9 (September 1999): 209–24.

Singer, Peter. *Rethinking Life and Death*. New York: St. Martin's Griffin, 1994.

Surber, Jere. "Obligations to Future Generations: Explorations and Problemata." *The Journal of Value Inquiry* 11 (Summer 1977): 104–16.

———. "Kant, Levinas, and the Thought of the Other." *Philosophy Today* (Fall 1994): 294–316.

———. "Ethical Principles. Levinas and the Tradition." Denver University 1996.

The Human Genome, Nature 409 (15 January 2001): 745–964.

The Human Genome, Science 291 (17 February 2001): 1145–1434.

UNESCO. "General Conference of UNESCO Adopts Universal Declaration on the Human Genome and Human Rights and a Resolution for its Implementation." *International Digest of Health Legislation* 49 (1998): 417–21.

Walters, Leroy in Craig Donegan, "Gene Therapy's Future." *The Congressional Quarterly Researcher* 5 (8 December 1995): 1089–112.

Watson, James. "The Human Genome Project: Past, Present, and Future." *Science* 248 (6 April 1990): 44–48.

Wivel, Nelson A. and Leroy Walters. "Germ Line Gene Modification and Disease Prevention: Some Medical and Ethical Perspectives." *Science* 262 (22 October 1993): 533–37.

Zanjani, Esmail and W. French Anderson. "Prospects for *in Utero* Human Gene Therapy." *Science* 285 (2000): 2084–96.

Zuk, P. A., et al., "Multilineage Cells from Human Adipose Tissue: Implications for Cell-Based Therapies." *Tissue Engineering* 7 (2001): 211–28.

Index

About the Author

Marilyn E. Coors is assistant professor at the Center for Bioethics and Humanities at the University of Colorado Health Sciences Center. In addition, she writes and speaks nationally about the ethics of human genetics. She is also clinically involved in ethics consultations and conducts research with human subjects on informed consent in genetic testing. Coors received her undergraduate education at Cornell University where she majored in biological science. She then attended Denver University, earning a M.S. in cytogenetics, a M.A. in ethics and a Ph.D. in bioethics. In addition, she serves as a community volunteer and with her husband, Peter, has raised six children in Golden, Colorado.